Intelligent Document Processing with AWS AI/ML

A comprehensive guide to building IDP pipelines with applications across industries

Sonali Sahu

BIRMINGHAM—MUMBAI

Intelligent Document Processing with AWS AI/ML

Publishing Product Manager: Dhruv Jagdish Kataria

Content Development Editor: Priyanka Soam

Technical Editor: Sweety Pagaria

Copy Editor: Safis Editing

Project Coordinator: Farheen Fathima

Proofreader: Safis Editing

Indexer: Rekha Nair

Production Designer: Joshua Misquitta

Marketing Coordinator: Shifa Ansari

First published: October 2022

Production reference: 1300922

Published by Packt Publishing Ltd.

Livery Place

35 Livery Street

Birmingham

B3 2PB, UK.

ISBN 978-1-80181-056-2

www.packt.com

For my parents, for always loving and believing in me.

For all the women who can do whatever they want, when they want – or will one day.

Contributors

About the authors

Sonali Sahu is a leading Intelligent Document Processing **Artificial Intelligence (AI)** and **Machine Learning (ML)** solutions architect on the team at Amazon Web Services. She is a passionate technophile and enjoys working with customers to solve complex problems using innovation. Her core area of focus is AI and ML. She has both breadth and depth of experience working with technology, with industry expertise in healthcare and insurance. She has significant architecture and management experience in delivering large-scale programs across various industries and platforms.

About the reviewer

Winnie Tung has over 30 years of experience solving some of the world's most difficult technical problems in the financial services industry. She is currently modernizing the AI/ML platform at JPMC. Before that, she worked in AWS Professional Services, specializing in developing AI/ML solutions for the real world. She helps customers to operationalize and manage AI/ML solutions at scale.

Table of Contents

3

Accurate Document Extraction with Amazon Textract   45

4

Accurate Extraction with Amazon Comprehend   67

Part 2: Enrichment of Data and Post-Processing of Data

5

6

7

Part 3: Intelligent Document Processing in Industry Use Cases

8

9

10

Preface

The volume of documents is exponentially growing in the digital era, and it has become paramount to process this data accurately and in an accelerated manner to get value out of it. Most often, the data is in raw document format, and being able to process these documents in an accelerated manner is critical to meeting the growing business needs, but legacy document processing doesn't meet this growing demand.

This book is a comprehensive guide that takes you through the fundamentals of **Artificial Intelligence (AI)** and **Machine Learning (ML)** and the core concepts required to process any type of document. You will also obtain hands-on experience with popular Python libraries for automating document processing. This book not only starts with the basics but also takes you through real industry use cases – for document processing in the healthcare industry to deliver value-based care, for claims processing in the insurance industry, and for accelerating loan application processing in the financial industry. That way, you are learning how to apply your skill sets to practical problems.

By the end of this book, you will have mastered the fundamentals of document processing with ML using practical implementations.

Who this book is for

This book is for technical professionals and thought leaders who want to understand and solve business problems by leveraging insights from their documents. If you want to learn about ML and AI and solve real-world use cases, such as document processing with technology, this book is for you. In order to learn from this book, you should have a basic knowledge of AI, ML, and Python programming concepts. The book is also excellent for developers who want to explore AI/ML with industry use cases.

What this book covers

Chapter 1, *Intelligent Document Processing with AWS AI and ML*, will explain how AWS wants to make ML accessible to everyone. For that reason, it has defined a three-layer AWS ML stack. AWS AI services can be called and leveraged by calling an API. First, the reader will learn about the AWS AI/ML stack. Then, we will define document processing, the challenges in document processing, and how AWS can help. We will also discuss common IDP use cases across industries. Finally, we will show the reader the stages of the IDP pipeline.

Chapter 2, Document Capture and Categorization, will detail how to collect data in a scalable, highly available data store. We will look into some of the security features for our data capture stage. Then, we will look into the accurate classification of documents. Readers will learn about the document splitter and how to use it on a code sample. Readers will learn to train their custom classifiers to accurately classify their document types.

Chapter 3, Accurate Document Extraction with Amazon Textract, will dive into key use cases for extracting data accurately from structured, unstructured, and semi-structured types of documents. Readers will learn about specialized documents, such as invoices, receipts, driver's licenses, and passports, and how we can leverage the AWS AI service Amazon Textract for accurate extraction.

Chapter 4, Accurate Extraction with Amazon Comprehend, will explain document extraction with Amazon Comprehend. Here, we will learn about the extraction features for Entities and Custom Entities in Amazon Comprehend. Readers will learn how to train their own custom Comprehend model with Amazon Comprehend. Finally, the reader will learn about the key phrases to extract for accurate document tagging and categorization.

Chapter 5, Document Enrichment in Intelligent Document Processing, will explore the document enrichment stage of IDP. Readers will learn about document enrichment and the redaction of sensitive information with PII detection in Amazon Comprehend. They will learn about extracting health insights from Amazon Comprehend Medical and how we can augment document processing with health insights and ontology linking.

Chapter 6, Review and Verification of Intelligent Document Processing, will elaborate on the post-processing stage, with completeness checks and access control. Readers will learn about the document completeness check during the post-processing of a document. They will also learn about PII detection in Comprehend and PHI detection in Comprehend Medical, with APIs for sensitive data redaction, and setting policies for right access control. Finally, the reader will learn about accuracy checks with human review.

Chapter 7, Accurate Extraction and Health Insights with Amazon HealthLake, will start with a brief introduction to healthcare interoperability with FHIR and explain the requirement to store documents in a healthcare datastore, which can be done with Amazon HealthLake. Readers will learn about the features of Amazon HealthLake and how to extend IDP to process and store documents in the health datastore.

Chapter 8, IDP Healthcare Industry Use Cases, will explore healthcare prior authorization and healthcare claims processing as IDP use cases. Readers will learn about the prior authorization process and how to build an IDP pipeline for prior authorization to accelerate the pre-certification process. Finally, the reader will learn about the claims adjudication process and build an end-to-end IDP pipeline for it.

Chapter 9, *Intelligent Document Processing – Insurance Industry*, will look into two use cases in the insurance industry – processing benefit registration and claims adjudication – as IDP solutions. Readers will learn how to use the various stages in the IDP pipeline to build and automate these use cases. Finally, we will accurately extract data from multiple document types and layouts for the verification of the claims form.

Chapter 10, *Intelligent Document Processing – Mortgage Processing*, will analyze lending document processing as an IDP solution. Readers will learn about mortgage and lending document processing with the IDP pipeline. Finally, we will accurately extract data from multiple document types and layouts for the verification of mortgage documents.

To get the most out of this book

You will need access to an AWS account, so before getting started, we recommend that you create one.

Software/hardware covered in the book	Operating system requirements
Sign up with an AWS account	Windows OS
Access to a web browser to access AWS	
Python	

If you are using the digital version of this book, we advise you to type the code yourself or access the code from the book's GitHub repository (a link is available in the next section). Doing so will help you avoid any potential errors related to the copying and pasting of code.

Download the example code files

You can download the example code files for this book from GitHub at `https://github.com/PacktPublishing/Intelligent-Document-Processing-with-AWS-AI-ML-`. If there's an update to the code, it will be updated in the GitHub repository.

We also have other code bundles from our rich catalog of books and videos available at `https://github.com/PacktPublishing/`. Check them out!

Download the color images

We also provide a PDF file that has color images of the screenshots/diagrams used in this book. You can download it here: `https://packt.link/2mHlD`.

Conventions used

There are a number of text conventions used throughout this book.

`Code in text`: Indicates code words in text, database table names, folder names, filenames, file extensions, pathnames, dummy URLs, user input, and Twitter handles. Here is an example: "Comprehend can take time to train your model. You can use Amazon Comprehend's `describe_document_classifier()` command or check on the AWS Management Console for the completion status."

A block of code is set as follows:

```
chapter2_syncdensedoc = "syncdensetext.png"
display(Image(url=s3.generate_presigned_url('get_
object', Params={'Bucket': s3BucketName, 'Key': chapter2_
syncdensedoc})))
```

Bold: Indicates a new term, an important word, or words that you see onscreen. For instance, words in menus or dialog boxes appear in **bold**. Here is an example: "The installer will present the following **License Agreement** screen. Click **I Agree**."

> **Tips or Important Notes**
> Appear like this.

Get in touch

Feedback from our readers is always welcome.

General feedback: If you have questions about any aspect of this book, mention the book title in the subject of your message and email us at `customercare@packtpub.com`.

Errata: Although we have taken every care to ensure the accuracy of our content, mistakes do happen. If you have found a mistake in this book, we would be grateful if you would report this to us. Please visit `www.packtpub.com/support/errata`, selecting your book, clicking on the Errata Submission Form link, and entering the details.

Piracy: If you come across any illegal copies of our works in any form on the Internet, we would be grateful if you would provide us with the location address or website name. Please contact us at `copyright@packt.com` with a link to the material.

If you are interested in becoming an author: If there is a topic that you have expertise in and you are interested in either writing or contributing to a book, please visit `authors.packtpub.com`.

Share your thoughts

Once you've read *Intelligent Document Processing with AWS AI/ML*, we'd love to hear your thoughts!
Scan the QR code below to go straight to the Amazon review page for this book and share your feedback.

https://packt.link/r/1-801-81056-7

Your review is important to us and the tech community and will help us make sure we're delivering
excellent quality content.

Part 1:
Accurate Extraction of
Documents and Categorization

In the first part, we will start with a brief introduction to **Intelligent Document Processing** (**IDP**) with AWS AI and ML, and then you will learn about accurate custom classification in the IDP pipeline. Next, you will learn about accurate data capture and data extraction in the IDP pipeline. Finally, the focus will be on Amazon Comprehend entities and the custom entities feature to leverage during the enrichment stage of the IDP pipeline.

This section comprises the following chapters:

1
Intelligent Document Processing with AWS AI and ML

It was a Wednesday evening – I was busy collecting all my receipts and filling out my insurance claim document. I wanted my health insurance to provide reimbursement for the COVID-19 test kits that I had purchased. The next day, I went to the post office to send the documents through postal mail to my insurance provider. This made me think how we are still working with physical documents in the 21st century. With my approximate math, this month alone, we will use 650 million documents per month, considering that 2% of the entire US population buys a test kit and applies for reimbursement using a paper-based application. This is a ton of documents in this instance. In addition to physical copies, we may have tons of documents that might just be scanned documents – we are looking at manual processing for these documents too. Can we do any better in the 21st century to automate the processing of these documents?

Besides this particular instance, we use documents for many other use cases across industries, such as claims processing in the insurance industry, loan, and mortgage documents in the financial industry, and legal and contract documents. If you have bought a house or refinanced a house, you will already be aware of the number of documents that you need to use for loan processing. IDC predicts worldwide data to exceed 175 zettabytes by 2025. The volume of data is huge. On top of the volume of data, we are talking about data of different formats and unstructured – some are forms, as with insurance claims, and some can be dense text, as with legal contractual documents. The volume and varying formats of documents make manual processing time-consuming, error-prone, and expensive. According to IDC, there is a 23% growth in data every year. The immense scale and format of documents make it a challenge to process them. Moreover, the legacy or traditional document extraction technologies can work well for pristine documents, but when document quality varies, the performance of those early-generation systems frequently does not meet customer needs. Manual document extraction carried out by a human workforce introduces variability into the process since people make mistakes and double-checking all work is not cost-effective. The most important of these factors is the ability to get the key information from the documents into your decision-making systems to make high-quality decisions more quickly and based on accurate information. Hence, we are all looking for efficient, less time-consuming, cost-effective ways to process our documents for better insights.

In this introductory chapter, we will be establishing the basic context to familiarize you with some of the underlying concepts of document processing, the challenges in document processing, and how AWS **Artificial Intelligence (AI)/Machine Learning (ML)** services can help solve these problems.

We will be covering the following topics in this chapter:

- Understanding common document processing use cases across industries
- Understanding the AWS ML and AI stack
- Introducing Intelligent Document Processing pipeline

Understanding common document processing use cases across industries

We started with a simple claims processing use case in the healthcare industry. But document processing challenges occur across multiple use cases and industries. For example, with a single patient generating nearly 80 megabytes of data each year in imaging and **Electronic Medical Record (EMR)** data, according to 2017 estimates, RBC Capital Markets projects that *by 2025, the compound annual growth rate of data for healthcare will reach 36%*. When a patient visits a physician, an immense amount of data is generated. Equally, when you speak with customers, they say they have petabytes of data in their archive, which is sitting there in a drive or tape drive without being processed further for legal or regulatory reasons, and most of it is unstructured data. For example, some healthcare providers in the US store medical history records for at least 7 years as per the regulation. If we can analyze a patient's historical data, we can build a predictive model for any chronic disease. This data is a gold mine, but because of the lack of an efficient, cost-effective mechanism for document processing, it sits there unused. Most of this data is currently stored as archived data and retired after the 7-year period is over. Can we use this data to derive insights for better healthcare outcomes?

Similarly, in the financial industry, there is a need for document processing – for example, when processing mortgage documents. Anyone who has bought a new home or refinanced their home must know the number of documents and different document types that we deal with for mortgage processing. McKinsey's report emphasizes that mortgage providers should get things right the first time to reduce any delay in processing. To address the timely verification of these documents, we need to empower loan officers with the right tools, automation, and insights. The immense volume and format of documents and the need to derive insights from them require automation with the right indexing, categorization, and extraction, with human reviews as needed to detect anomalies and get the mortgage documents right the first time for timely processing.

It is not only the healthcare or financial industries that require document processing but also industries across verticals and use cases such as legal documents and contracts, insurance, ID handling, and enrollments with the use of advanced technologies such as *AI and ML*, wants to automate document processing with advanced AI and ML technologies. Intelligent Document Processing uses AI-powered automation and ML to classify, extract, transform, and enrich our documents for consumption. Before discussing advanced technologies and solutions, it is always good to start with the basics. So, let's first set the foundation of AI and ML.

Understanding the AWS ML and AI stack

Just five decades ago, ML was still a thing of science fiction. But today, it is proven to be an integral part of our everyday lives. It helps us drive our cars, recommends personalized shopping experiences, and helps us utilize voice-enabled technologies such as Alexa. The early days of AI and ML began with simple calculators or chessboard games but by the 20th century, this has evolved into diagnosing cancer and more. The initial theory of ML was in research and labs and now it has moved from labs to real lives applications across industries. This is a change in the adoption of AI and ML.

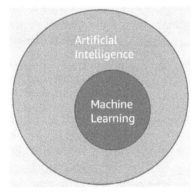

Figure 1.1 – AI and ML

What is AI? AI is a wide range of computer science branches related to building smart machines. And ML is a subset or application of AI, as shown in *Figure 1.1*. The goal of ML is to let the machine learn automatically without any programming or human assistance. We want the machine to learn from its own experience and provide results. You gather data and the model learns and corrects itself based on this data. One of the famous historical achievements of AI or ML is Alan Turing's paper and the subsequent development of the Turing Test in the 1950s. This established the fundamental goal and vision for AI. This focused on one main thing – can machines learn like humans? After 2 years, Arthur Samuels, another pioneer in the **computer science** and gaming industry, wrote the very first computer learning program for playing the game checkers. It was programmed to learn from the moves that allowed it to win and then program itself to play the game. With some of the recent AI and ML accomplishments, in the year 2015, AWS launched its own ML platform to make its models and ML infrastructure more accessible.

Now, we see AI and ML in our everyday usage. If you have used any e-commerce or online media or entertainment platforms, you must be familiar with receiving personalized recommendations or using conversational chatbots and virtual assistance with AI services. These personalized recommendations and experiences drive user engagement. Similarly, any helpdesk calls at contact centers can be automated with AI, driven to reduce the burden on human beings with reduced costs. Moreover, AI can be used in automatic document processing for accurate extraction and analysis and to instantly derive insights from it, as in loan processing or claims processing.

Now, we see a wide presence of ML and AI in our everyday usage and industries are busy building newer models to learn better and more quickly to give accurate predictions and accelerate business value. But the main question is – can we share the experience and knowledge that we learned when building models? Can a builder re-use an already trained model for its own business without spending time and effort to train another model? So, can we share our experience and knowledge and ML models for any builder to use and focus on their business needs?

The answer is yes, and for that reason, AWS has divided its ML stack into three broad categories. Let's discuss the three individual AI/ML stacks in detail and their core goals in solving user requirements in the following figure:

ML Frameworks	PyTorch, Apache MXNet, TensorFlow, Hugging Face
Infrastructure	Amazon EC2, GPU/CPU, Inferentia, Trainium, Habana Gaudi, FPGA, Elastic Inference

Figure 1.2 – The ML framework and infrastructure at the bottom of the AWS stack

At the bottom of the AWS AI or ML stack, we see services and features targeted at expert ML practitioners who have the expertise and are comfortable working with ML frameworks, algorithms, and deploying their ML infrastructure. Some of the AWS ML frameworks and infrastructure are shown in *Figure 1.2*. AWS offers users their framework of choice, thus supporting ML frameworks such as **PyTorch**, **Apache MxNet**, and **TensorFlow** to run optimally on the AWS platform. The bottom layer also stacks **CPU** and **GPU** instances. Decades ago, obtaining GPU resources to accelerate your ML workload was a wild dream for general ML builders. You might have to reach out to a supercomputing center to get ahold of GPU resources. But today, you can access GPUs at your fingertips with AWS. AWS gives users the option to customize and select instances with customized memory, vCPU, architectures, and more. AWS added Trainium, a second ML chip optimized for deep learning training.

Not only that, but to democratize the ML infrastructure, AWS offers Inferentia to drive high-performance deep learning inference on the cloud at a fraction of the cost:

SageMaker Canvas	No-Code ML
SageMaker Studio Lab	Learn ML
SageMaker Studio IDE/RSTUDIO	ML with CI/CD

Figure 1.3 – ML services in the middle of the AWS stack

The middle layer in the AI or ML stack is more targeted toward an ML builder who wants to build, train, and deploy their own ML models. Some of the AWS offerings for ML services are shown in *Figure 1.3*. This layer makes ML more accessible and expansive. Amazon SageMaker helps data scientists and developers prepare, build, train, and deploy high-quality ML models quickly by bringing together a broad set of capabilities purpose-built for ML. Amazon SageMaker offers JumpStart to help you quickly get started with a solution by automatically extracting, processing, and analyzing documents for accelerated and accurate business outcomes. It offers an integrated Jupyter notebook for authoring your model with pre-built optimized algorithms. But at the same time, it gives options to ML users to bring their own algorithms and frameworks. It offers a managed, scalable, and secure training and deployment platform for your ML process. To learn more about Amazon SageMaker, you can also refer to the book *The Machine Learning Solutions Architect* written by *David Ping* and published by Packt.

You can find this book here: `https://www.packtpub.com/product/the-machine-learning-solutions-architect-handbook/9781801072168`.

Use case	AWS AI services
Personalization	Amazon Personalize
Intelligent Search	Amazon Kendra
Intelligent Document Processing	Amazon Comprehend Medical Amazon Comprehend Amazon Textract
Call Center Modernization	Amazon Transcribe Amazon Comprehend Amazon Lex
Content Moderation	Amazon Rekognition
Industrial	Amazon Monitron Amazon Lookout for Equipment Amazon Lookout for Vision
Healthcare	Amazon Comprehend Medical Amazon Transcribe Medical Amazon HealthLake

Figure 1.4 – AI services at the top of the AWS stack

AWS designed the top layer to put ML in the hands of every single developer. These are AI services. AWS drew on its experience with Amazon.com and its ML services to offer highly accurate, API-driven AI services. You do not need to be an ML expert to call on the pre-trained models leveraging APIs. Rather, you can use AI services to enhance your customer experience, improve productivity, and get a faster time to market with ready-made ML models. At the core of the AI services, we have **Vision** services, with **Amazon Rekognition** and **AWS Panorama**. For **Speech**, we have services such as **Amazon Polly, Amazon Transcribe**, and **Call Analytics**; and for chatbots, **Amazon Lex**. You can leverage these speech and bot services for use cases such as call center modernization. For leveraging the experience of Amazon.com on a recommendation system, it offers **Amazon Personalize**. In this book, we will dive deep into the document processing use cases with its **text and document services** such as **Amazon Textract** and **Amazon Comprehend**. To help the customer with industry-specific use cases, AWS AI services are also categorized in terms of industrial use, with AWS AI services such as **Monitron** and **Lookout**, and healthcare technologies, with AI services such as **Amazon Comprehend Medical, HealthLake**, and **Transcribe Medical**. In *Figure 1.4* here, we are showing how AI services can be aligned to specific industry use cases. But in this book, we will dive deeper into IDP use cases in particular.

Some of the main benefits of AWS AI services are that the models are fully managed and AWS takes care of the undifferentiated heavy lifting in building, maintaining, patching, or upgrading servers or hardware required for the model(s) to run. You can customize and interact with the AI models and perform predictions via API calls or directly from the AWS console. AWS AI services enable you to have performant and scalable solutions with serverless technologies, which can be called using these AI service APIs. You can just call APIs using a serverless architecture that scales automatically as the document processing demand grows or shrinks. This is highly performant, with low latency and timely delivery of your business use case:

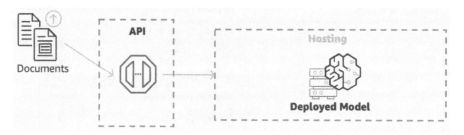

Figure 1.5 – Accessing AWS AI services with an API call

With AWS AI or ML offerings, we have multiple technologies available to implement the same use case. There are trade-offs when using AI services that are API-driven versus ML services. We will dive deeper into the comparison of AI and ML models for IDP in *Chapter 3, Accurate Document Extraction with Amazon Textract*, under the *Introduction to Textract* section.

Alright, it's time to get started with an overview of IDP. Now that we understand how AWS cloud infrastructure and services will help us accelerate our AI or ML workload, let's dive into the IDP pipeline and its applications across industries.

Introducing Intelligent Document Processing pipeline

IDP seems simple but in reality, it is a complex challenge to solve. Imagine a physical library – racks and racks of books divided and arranged in rows tagged with the right author and genre. Have you wondered about the human workforce behind doing this diligent, structured work to help us find the right book in a timely and efficient manner?

Similarly, as you know, we deal with documents across industries for various use cases. In the traditional world, you would need many teams to go through the entire list of documents and manually read documents individually. Then, they would need to identify the category the document belongs to and tag it with the right keywords and topics so that it can be easily identifiable or searchable. Following the process, your main goal is to extract insights from these documents. This is a massive process and takes months and years to set up based on the volume of the data and the skill level of the manual workforce. Manual operations can be time-consuming, error-prone, and expensive. To onboard a new document type and update or remove a document type, these steps need to be followed incrementally. This is a significant investment, effort, and a lot of pressure on the manual workforce. Sometimes, the time and effort needed are not budgeted for and can cause significant delays or pause the process. To automate this, we need **digitization** and **digitalization**.

Digitization is the conversion of data to digital format. Digitization has to happen before digitalization. Digitalization is the process of using these digital copies to derive context-based insights and transform the process. After transformation, there must be a way to consume the information. This entire process is known as **the IDP pipeline**. Go through the following in *Figure 1.6* to get a detailed view of the IDP pipeline and its stages:

Figure 1.6 – The IDP pipeline and its stages

Now that we know what the IDP pipeline is, let's understand each phase of IDP in detail.

Data capture

In our library books example, we can go to a library directly, look for books, borrow a book, return a book, or just sit and read a book in the library. We want a place where all books are available, well-organized, easily accessible when we need them, and affordable. Similarly, at the data capture stage, documents are similar to our library books. During this stage, we collect and aggregate all our data in a secure, centralized, scalable data store. While building the data capture stage for your IDP pipeline, you have to take data sources, data formats, and the data store into consideration:

- **Document sources**: Data can come from various sources. It can be as simple as mobile capture, such as submitting receipts for reimbursement or submitting digital pictures of all your applications, transcripts, ID documents, and supporting documents during any registration process. Other sources can be simple fax or mail attachments.

- **Document format**: The data we speak about comes in different formats and sizes. Some can just be a single page, such as a driver's license or insurance card, but others can be multiple pages, such as in a loan mortgage application or with insurance benefit summary documents. But we categorize data into three broad categories: **structured**, **semi-structured**, and **unstructured**. Structured documents have structured elements in them, such as table-type elements. Unstructured documents have dense text, as in legal and contractual documents. Finally, semi-structured documents contain key-value elements, as in an insurance application form. But most often documents can have multiple pages with all the different category (structured, semi-structured, and unstructured) elements in them. There are also different types of digital documents – some can be image-based, with JPEGs and PNGs, and others can be PDF or TIFF types of documents with varying resolutions and printing angles.

- **Document store**: To store the untransformed and transformed documents, we need a secure data store. At times, we have a requirement to store metadata about documents, such as the author or date, which is not mentioned in the document, for future mapping of metadata to extraction results. Industries such as healthcare, the public sector, and finance should be able to store their documents and their results securely, following their security, governance, and compliance requirements. For easier, instantaneous, and highly performant access, they need storage with industry-leading easy-to-use management and simpler access from anywhere at the click of a button. The volume of data and documents is vast. To support it, we require a scalable data store, which can scale as per our needs. Another important factor is the high reliability and availability of your data store so that you can access it whenever you have a need. Moreover, given the high volume of documents, we are looking for a cost-effective document store.

Let's now move on to the next IDP phase.

Document classification

Going back to our book library example, the books are categorized and stacked by category. For example, for any fiction or non-fiction books, you can directly check the label on the rack and go to the section where you can find the books related to that category. Each section can be further subdivided into sub-sections or can be arranged by the first letter of the author's name. Can you imagine how difficult it would be to locate a book if it were not categorized correctly?

Similarly, at times, you receive a package of documents, or sometimes a single PDF, with all the required documents merged. A human can preview the documents to categorize them into their specified folder. This helps later with information extraction and metadata extraction from a variety of complex documents, depending on the document type. This process of categorizing the documents is known as **document classification** or a **document splitter**.

This process is crucial when we try to automate our document extraction process and when we receive multiple documents and don't have a clear way to identify each document type. If you are dealing with a single document type or have an identifiable way to locate the document, then you can skip this step in the IDP pipeline. Otherwise, classify those documents correctly before proceeding in the IDP pipeline.

Document extraction

Again, analogous to our library books, now that all the books are accurately categorized and stacked, we can easily find a book of our choice. When we read a book, we might come across multiple different formats of text, such as dense paragraphs interweaved between tables and some structured or semi-structured elements such as key values. As human beings, we can read and process that information. Human beings know how to read a table or key-value types of elements in addition to a paragraph of text. Can we automate this step? Can we ask a machine to do the extraction for us?

The process of accurately extracting all elements, including structural elements, is broadly known as **document extraction** in the IDP pipeline. This helps us to identify key information from documents through extensive, accurate extraction. The intelligent capture of the data elements from documents during the extraction phase of the IDP pipeline helps us derive new insights in an automated manner.

Some of the examples of the extraction stage include **Named Entity Recognition** (**NER**) for automatically identifying entities in unstructured data. We will look into the details more deeply in *Chapter 4, Accurate Extraction with Amazon Comprehend.*

Document enrichment

To get insights and business value out of your document, you will need to understand the dynamic topics and document attributes in your document. During the document enrichment stage, you append or enhance the existing data with additional business or domain-specific context from additional sources.

For example, while processing healthcare claims, at times, we need to refer to a doctor's note to verify the medical condition mentioned in the claims form. Additional documents such as doctor's notes are requested for further processing. We get a raw doctor's note deriving medical information such as details about medication and medical conditions – being able to get this directly from the main document is critical to enable business value such as improving patient care. To achieve this, we need the medical context, metadata, attributes, and domain-specific knowledge. This is an example of the enrichment stage of the IDP pipeline.

While entity recognition can extract metadata from texts of various document types, we need a process to recognize the non-text elements in our documents. This is where the object detection feature comes in handy. This can be extended further into identifying personal information with **Personally Identifiable Information** (**PII**) and **Protected Health Information** (**PHI**) detection methods. We can also de-identify our PII or PHI for further downstream processing. We will look into the details in *Chapter 5, Document Enrichment in Intelligent Document Processing*.

Document post-processing (review and verification)

Going back to our book library example, there are certain instances when a library gets a new book and places the book in a *new book* section instead of categorizing it by genre. These are some specific rules that we follow for certain books in that library. Some specific rules and post-processing are required to organize our books in the library.

Similarly, with document processing, you might want to use your business rules or domain-specific validation to check for its completeness. For example, during a claims processing pipeline in the insurance industry, you want to validate for insurer ID and additional basic information. This is to check for the completeness of the claims form. This is a type of post-processing in the IDP pipeline.

Additionally, the extraction process or the enrichment steps previously discussed may not be able to give you the accuracy required for your business needs. You may want to include a human workforce for manual review for higher accuracy. Having a human being review some or certain fields of your documents in an automated way for higher accuracy can also be a part of the post-processing phase in IDP. Human review can be expensive, so in an automated manner, we will only process limited data on our documents in this way as per our business needs and requirements. We will further discuss this phase in *Chapter 6, Review and Verification of Intelligent Document Processing*.

Consumption

In our book library example, we always wish for a centralized, unified portal to track all our library books and their statuses. To maintain a digital library or online library system, nowadays, libraries support online catalogs where you can check all the books in a library in a centralized portal and their reservation statuses, the number of copies, and additional book information about its author and ratings. This is an example where we are to not only maintain and organize a book library but also integrate library information with our portals. We might maintain multiple different portals or tracking systems to manage and maintain our library books. This is an example of the integration and consumption stage for our library books.

Similarly, in our IDP pipeline, we collect our documents and categorize them during the data capture and classification stages. Then, we accurately extract all the required information from our documents. With the **enrichment** stage, we derived additional information and transformed our data for our business needs. Now is the time to consume the information for our business requirements. We need to integrate with our existing systems to consume the information and insights derived from our documents. Most of the time, I come across customers already using an existing portal or tracking system and wanting to integrate the insights derived from their documents with the existing system. This will also help them to build a 360 view of their product from the consumer perspective. At other times, the customer wants just a data dump in their database for better, faster queries. There can be many different ways and channels you want to use to consume the extracted information. This stage is known as the **consumption** or **integration stage** in our IDP pipeline.

Let's now summarize the chapter.

Summary

In this chapter, we discussed the current challenges in document processing and how IDP can help overcome those challenges. We introduced IDP by tracing the origins of AI, how it has evolved over the last few decades, and how AI became an integral part of our everyday lives.

We then reviewed industry trends and market segmentation and saw with examples how important it is to automate document processing. We also discussed IDP across industry use cases. We read an example of how patient data can be collected and enriched to better patient outcome prediction.

Finally, we reviewed the stages of the IDP pipeline such as data capture, data classification, data extraction, data enrichment, and data post-processing. This chapter gave readers an understanding of IDP and the various stages involved to automate the end-to-end pipeline.

In the next chapter, we will go through the details of the data capture stage and document classification with AWS AI services. We will also look into the details of AWS AI services such as Amazon Comprehend custom classification and Amazon Rekognition for document classification.

References

- *IDC Report*: https://www.idc.com/getdoc.jsp?containerId=prUS47560321

2
Document Capture and Categorization

One of the first stages of an **Intelligent Document Processing** (**IDP**) pipeline is to collect your documents and store them in a highly available, reliable, and secure data store. Data is our gold mine, and to extract insights from our documents, we need to understand our data and pre-process it as needed. Most of the time, organizations receive a package of documents that are not labeled. To understand the documents, you need to manually scan these documents and label them into the right category, which is known as the document classification stage of the IDP pipeline. Thus, we are looking for an automated process for data collection and document classification.

In this chapter, we will be covering the following topics:

- Understanding data capture with Amazon S3
- Understanding document classification with Amazon Comprehend's custom classifier
- Understanding document categorization with computer vision

Technical requirements

To do the hands-on labs in the book, you will need access to an AWS account at `https://aws.amazon.com/console/`. Please refer to the *Signing up for an AWS account* subsection within the *Setting up your AWS environment* section for detailed instructions on how you can sign up for an AWS account and sign in to the AWS Management Console. The Python code and sample datasets for the solution discussed in this chapter are available at `https://github.com/PacktPublishing/Intelligent-Document-Processing-with-AWS-AI-ML-/tree/main/chapter-2`.

Signing up for an AWS account

In this chapter and all the subsequent chapters in which we will run code examples, you will need access to an AWS account. Before getting started, we recommend that you create an AWS account.

> **Important note**
> Please use the AWS Free Tier, which enables you to try services free of charge based on certain time limits or service usage limits. For more details, please see `https://aws.amazon.com/free`.

Setting up your AWS environment

Please go through the following steps to create an AWS account:

1. Open `https://portal.aws.amazon.com/billing/signup`.

2. Click on the **Create a new AWS account** button at the bottom left of the page.

3. Enter your email address and a password, confirm the password, and provide an AWS account name (this can be a reference to how you will use this account, such as sandbox).

4. Select the usage type (**Business** or **Personal**), provide your contact information, read and agree to the terms of the AWS Customer Agreement, and click **Continue**.

5. Provide credit card information and a billing address, and click **Continue**.

6. Go through the rest of the steps to complete your AWS account signup process. Please make a note of your user ID and password; this is the *root* access to your AWS account.

7. Once the AWS account is created, go to the AWS Management Console (`console.aws.amazon.com`) and sign in using the root credentials you created in the previous steps.

8. Type `IAM` in the services search bar at the top of the console and select **IAM** to navigate to the IAM console. Select **Users** from the left pane in the IAM console and click on **Add User**.

9. Provide a username, and then select **Programmatic access** and **AWS Management Console access** for **Access Type**. Keep the password as **Autogenerated** and keep **Require password reset** selected.

10. Click **Next: Permissions**. On the **Set Permissions** page, click on **Attach existing policies** directly and select the checkbox to the left of **Administrator Access**. Click **Next** twice to go to the **Review** page. Click **Create user**.

11. Now, go back to the AWS Management Console (`console.aws.amazon.com`) and click **Sign In**. Provide the IAM username you created in the previous step and the temporary password, and then create a new password to log in to the console.

12. Log in to your AWS account when prompted in the various chapters and sections. You now have access to the AWS Management Console (`https://aws.amazon.com/console/`).

In the next section, we will show you how to create an S3 bucket and upload your documents.

Creating an Amazon SageMaker Jupyter notebook instance

In this section, we will see how to create a notebook instance in Amazon SageMaker. This is an important step, as most of our solution examples are run using notebooks. After the notebook is created, please follow the instructions to use the notebook in the specific chapters based on the solution being built.

Follow these steps to create an Amazon SageMaker Jupyter notebook instance:

1. Log in to the AWS Management Console if you haven't already. Type `SageMaker` in the services search bar at the top of the page, select **SageMaker** from the list, and click on it to go to the **Amazon SageMaker management** console.

2. In the **SageMaker** console, on the left pane, click on **Notebook** to expand the option, and then click **Notebook instances**.

3. On the **Notebook instances** page, click the **Create** notebook instance button at the top right.

4. Type a name for the notebook instance and select a suitable notebook instance type. For most of the solution builds in this book, an AWS Free Tier (`https://aws.amazon.com/free`) instance such as `ml.t2.medium` should suffice.

5. In the **Permissions and encryption** section, click the **IAM role** list and choose **Create a new role**, then choose **Any S3 bucket**, and then select **Create Role**.

6. Accept defaults for the rest of the fields and click **Create notebook instance**.

Your notebook instance will take a few minutes to be provisioned; once it's ready, the status will change to **In Service**. In the next few sections, we will walk through the steps required to modify the IAM role we attached to the notebook.

Changing IAM permissions and trust relationships for the Amazon SageMaker notebook execution role

In AWS, security is "job zero," and each service needs explicit role level access to call another service – for example, if we want to call a Textract service inside a SageMaker notebook instance, we need to give required permission to SageMaker Notebook, using **Identity and Access Management** (**IAM**) to call the Textract service. In this section, we will walk through the steps needed to add additional permission policies to our Amazon SageMaker Jupyter notebook role:

1. Open the Amazon SageMaker console by typing `sagemaker` in the services search bar at the top of the page in the AWS Management Console, and then select **Amazon SageMaker** from the list.

2. In the **Amazon SageMaker** console, on the left pane, expand the Notebook and click **Notebook instances**.

3. Click the name of the notebook instance you need to change permissions for.

4. On the **Notebook instance settings** page, scroll down to **Permissions and encryption** and click **IAM role ARN**.

5. This will open the IAM Management Console, and your role summary will be displayed along with the permissions and other details for your role. Click **Attach policies**.

6. On the **Add permissions** to **<your execution role name>** page, type `textract` in the search bar, select the checkbox next to the policy (`AmazonTextractFullAccess`) you are interested in, and click **Attach policy** at the bottom right. You should now see the policy attached to your role. Follow the same process for the **Comprehend**, **Comprehend medical**, **rekognition**, **healthLake**, and **amazon augmented AI** services. To simplify, and for the hands-on code, you can select **full access policies** for the respective services. But for the production workload, and to follow security guidelines, it is recommended that you select minimal access privilege

> **Important note**
> To run all the hands-on labs in this book, Full list of required IAM Roles can be found in `https://github.com/PacktPublishing/Intelligent-Document-Processing-with-AWS-AI-ML-/blob/main/IAM_Roles`.

7. In some cases, we may need a custom policy for our requirement rather than a managed policy provided by AWS. Specifically, we would add an inline policy (`https://docs.aws.amazon.com/IAM/latest/UserGuide/id_roles_use_passrole.html`) to allow **PassRole** of our SageMaker notebook execution role to services that can assume this role (added at the **Trust relationships** stage), for actions that need to be performed from the notebook. Click **Create inline policy** on the right of your SageMaker notebook execution role summary page.

8. Now, click the **JSON** tab and paste the following JSON statement into the input area:

```
{ "Version": "2012-10-17", "Statement": [ {
"Action": [
"iam:PassRole"
],
"Effect": "Allow",
"Resource": "<IAM ARN of your current SageMaker
notebook execution role>"
}
]
}
```

9. Click **Review policy**.

10. On the **Review policy** page, type a name for your policy and click **Create policy** at the bottom right of the page.

11. Now that you know how to attach permissions and an inline policy to your role, let's go to the last step of this section, updating trust relationships (`https://docs.aws.amazon.com/`

directoryservice/latest/admin-guide/edit_trust.html) for your role. On the **Summary** page for your SageMaker notebook execution role, click the **Trust relationships** tab, and click **Edit trust relationship**.

12. Copy the following JSON snippet and paste it into the **Policy Document** input field. This statement gives Amazon SageMaker, **Amazon Simple Storage Service** (**Amazon S3**), and Amazon Comprehend the ability to assume the SageMaker notebook execution role permissions. Depending on the chapter and the use case we are building, the services that will need to assume the role will vary, and you will be instructed accordingly. For now, please consider the following JSON snippet as an example to understand how to edit trust relationships:

```
{ "Version": "2012-10-17", "Statement": [
{ "Effect": "Allow",
 "Principal":
{ "Service":
[ "sagemaker.amazonaws.com",
  "s3.amazonaws.com",
  "comprehend.amazonaws.com" ]
},
"Action": "sts:AssumeRole"
}
]
}
```

13. Click the **Update Trust Policy** button at the bottom right of the page.

14. You should see the trusted entities updated for your role.

> **Important note**
>
> You cannot attach more than 10 managed policies to an IAM role. If your IAM role already has a managed policy from a previous chapter, please detach this policy before adding a new policy as per the requirements of your current chapter. When we create an Amazon SageMaker Jupyter notebook instance (as we did in the previous section), the default role creation step includes permissions for either an S3 bucket that you specify or any S3 bucket in your AWS account.

Now, you are all set.

Go to the GitHub link https://github.com/PacktPublishing/Intelligent-Document-Processing-with-AWS-AI-ML- and then the chapter-2.pynb notebook.

Now that we are done with the technical requirements, let's look at the components involved in the data capture stage.

Understanding data capture with Amazon S3

Document capture or ingestion is a process to aggregate all our data in a secure, centralized, scalable data store. While building a data capture stage for your IDP pipeline, you have to take data sources, data format, and a data store into consideration.

Data store

The first step is to store our documents for transformation. To store documents, we can use any type of document store, such as a local filesystem or **Amazon S3**. For this IDP pipeline, we will be leveraging AWS AI services, and we recommend, for an easier, more secure, and more scalable document store, to leverage Amazon S3, an object storage service that offers industry-leading scalability, data availability, security, and performance. Amazon S3 has 11 9s of durability, and millions of customers all around the world leverage Amazon S3 for their data store.

Many regulatory industries, such as GE Healthcare, use Amazon S3 for data storage during their digital transformation journeys. GE Health Cloud leveraged Amazon S3 to build a single portal for healthcare data access. Similarly, customers across industries such as the financial sector, public sectors, and so on have leveraged Amazon S3 for their secure data store. You can leverage Amazon S3 to build a data lake. You can run big data analytics and AI/ML applications to unlock insights from your documents. But, at times, you will see companies having their data source centralized on-premises, not wanting to store documents redundantly anywhere else for security reasons. We will discuss how to process documents from such data sources by leveraging AWS IDP in a later section. Moreover, you will see document processing requirements when data sources are in Salesforce, Workday, Box, SAP, and so on, and we can use a custom connector to process documents from these data sources, but connector implementation details are out of the scope of this book.

For this book, we will leverage Amazon S3 as a data store for IDP with AWS AI services. This offers ease of access and implementation for document processing with AWS AI services.

Data sources

Data can come from various different sources. Data capture can be synchronous where we need to process and get a result instantaneously. For example, we might want to submit receipts for reimbursement. We would need an instantaneous acknowledgment that our receipts are in the right resolution and have been uploaded successfully. We would not expect to receive an instant approval or denial result, which may at times need human reviews, but we expect an immediate acknowledgment.

I was speaking to one of our customers, who was working on building a smart portal for the online submission of university applications. As a process, you would expect the applicant needs to submit ID proof along with other details relevant to the application and the university enrollment application forms itself at the least. Moreover, different universities have their own university-specific application requirements. This process used to be manual and take a long time, with multiple iterations needed just to collect all required documents with the right resolutions. With the building of the smart portal, applicants could now immediately get an acknowledgment if all required documents with the right quality had been uploaded successfully or not. This can eliminate 2 weeks of additional time just to collect all required documents needed for the application. You can imagine this requirement with your use cases, where there is a need to process fully or partially documents to give instantaneous results. In *Figure 2.1*, you can see how various types of documents from different channels can be centralized in an Amazon S3 data store.

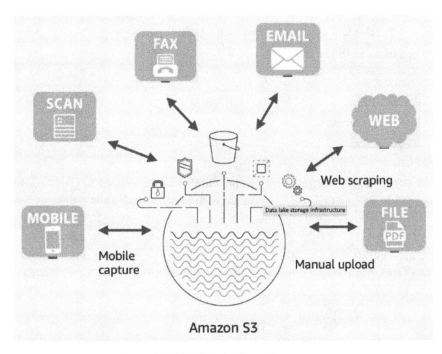

Figure 2.1 – The IDP pipeline – data capture

Let's see how to upload and process those documents for synchronous extraction using code. We will use the syncdensetext.png file for the following hands-on code.

We will be using the **AWS Boto3 library** to call into our AI service APIs. You may need to upgrade your boto3 libraries to get the latest updated access to APIs.

Refer to your notebook (https://github.com/PacktPublishing/Intelligent-Document-Processing-with-AWS-AI-ML-/blob/main/chapter-2/Chapter-2-datacapture.ipynb for full hands-on lab) section to import the boto3 library:

```
import boto3
from IPython.display import Image, display
!pip install textract-trp
```

```
from trp import Document
from PIL import Image as PImage, ImageDraw
import time
from IPython.display import IFrame
```

Follow these instructions to create an Amazon S3 bucket:

1. First, you get a handle for an S3 client by calling into the `boto3.client("S3")` API.

2. Create an S3 bucket and upload your sample documents.

3. Give a unique name to your bucket:

    ```
    # Create a Unique S3 bucket
    s3BucketName = <YOUR_BUCKET_NAME>
    print(s3BucketName)
    ```

> **Important note**
>
> Bucket names are globally unique, so be creative and give a unique name to your Amazon S3 bucket in <YOUR_BUCKET_NAME>. You might want to add a random hash number to your bucket name, as bucket names are globally unique.

4. Create an S3 bucket and upload the `syncdensetext.png` sample document file to your created S3 bucket. We will use this document as an example to process using the AWS AI service:

    ```
    # Upload document to your S3 Bucket
    print(s3BucketName)
    !aws s3api create-bucket --bucket {s3BucketName}
    !aws s3 cp syncdensetext.png s3://{s3BucketName}/
    syncdensetext.png
    ```

5. Now, display our `syncdensetext.png` document:

    ```
    chapter2_syncdensedoc = "syncdensetext.png"
    display(Image(url=s3.generate_presigned_url('get_
    object', Params={'Bucket': s3BucketName, 'Key': chapter2_
    syncdensedoc})))
    ```

The output for the preceding code looks like the following:

It was a Wednesday evening, and I was collecting all my receipts and busy filling out my insurance claim document. I wanted to submit to my Health Insurance for reimbursement for my COVID-19 test kits that I have purchased. The next day I went to the post office to send documents through postal mail to my insurance provider. This made me think, in the 21st century we are still working with physical documents. With my approximate math this month alone, we will get X number of documents per month considering 20% of entire US population buys a test kit. This is a tons of documents as in this instance. In addition to physical copies we have tons of documents which are might be just scanned documents. And we are looking for manual processing for these number of documents. Can we do any better in 21st century to automate the process of these documents?

Figure 2.2 – Sample document content

6. Once we have uploaded our sample document to our Amazon S3 bucket, we want to extract data elements from it. For that reason, we will get a "Textract" boto3 client:

```
# Amazon Textract client
textract = boto3.client('textract')
```

7. Then, we call Amazon Textract's detect_document_text API with our S3 bucket name and the document name.

8. We collect the Textract JSON response in the response object.

9. We loop through our JSON response to collect all the LINE and print it:

```
# Document Extraction with Amazon Textract
response = textract.detect_document_text(
    Document={
        'S3Object': {
            'Bucket': s3BucketName,
            'Name': chapter2_syncdensedoc
        }
    })
# Print All the lines
for line in response["Blocks"]:
    if line["BlockType"] == "LINE":
        print (line["Text"])
```

> **Important note**
> We will look into the API details of Amazon Textract in *Chapter 3, Accurate Document Extraction with Amazon Textract*.

Check the following output for all the extracted lines from the document.

```
It was a Wednesday evening, and I was collecting all my receipts and busy filling out my
insurance claim document. I wanted to submit to my Health Insurance for reimbursement for
my
COVID-19
test kits that I have purchased. The next day I went to the post office to send
documents through postal mail to my insurance provider. This made me think, in the 21° st
century we are still working with physical documents. With my approximate math this month
alone, we will get X number of documents per month considering 20% of entire US
population buys a test kit. This is a tons of documents as in this instance. In addition to
physical
copies we have tons of documents which are might be just scanned documents. And
we
are looking for manual processing for these number of documents. Can we do any better
in 21st century to automate the process of these documents?
```

Figure 2.3 – The LINES output of the synchronous API

Other data sources can be just simple fax or mail attachments or a bulk upload of documents. These documents need to be processed periodically or ad hoc to get insights from documents. Any asynchronous or batch processing operations can fit these use cases.

For example, for real estate insurance, the reviewer needs to check and validate real estate properties for accurate assessments. For that reason, they collect and upload insurance title documents from various sources and web and bulk upload to their data store. They manually extract the data elements from the documents such as address, lot number, and so on, and do the data entry in their system so that a user can easily find those data elements easily, which helps in making the right assessment. As you can see, in this use case, there is no need to process documents instantaneously, but you can asynchronously upload and extract elements from your documents.

Let's see how to upload and process those documents for asynchronous extraction in code.

Follow the steps mentioned previously to create your Amazon S3 bucket and upload the `sync_densetext.png` document to your S3 bucket.

This is the method to start a **Textract asynchronous** operation to extract elements from your Amazon S3 bucket:

1. We use the `boto3` client for Textract, and we call into the `start_document_text_detection` API of Amazon Textract to process our document's extraction asynchronously.

2. We pass in our bucket name and document name as parameters to the `start_document_text_detection` API.

3. We return the response for `JobId` from the method for asynchronous document processing:

```
def startasyncJob(s3BucketName, filename):
    response = None
    response = textract.start_document_text_detection(
```

```
        DocumentLocation={
            'S3Object': {
                'Bucket': s3BucketName,
                'Name': filename
            }
        })
        return response["JobId"]
```

The following is the method to check for the completion of Amazon Textract's asyncronous API extraction. The Textract asyncronous API takes time to process all our documents. Constant polling to check the status of the Textract processing may cause unnecessary CPU overhead, so we will use a sleep and pool mechanism to get the status of our asynchronous operation. There are additional options to pool status of an asynchronous API job. We will discuss this in detail in *Chapter 3, Accurate Document Extraction with Amazon Textract*:

4.   This is the function to get a result from our Textract asynchronous API, once the job is completed:

```
    def isAsyncJobComplete(jobId):
        response = textract.get_document_text_
    detection(JobId=jobId)
        status = response["JobStatus"]
        print("Job status: {}".format(status))
        while(status == "IN_PROGRESS"):
            time.sleep(10)
            response = textract.get_document_text_
    detection(JobId=jobId)
            status = response["JobStatus"]
            print("Job status: {}".format(status))
        return status
```

5.   We get the result in the `response` object and loop through it to pull all page blocks.

This function returns all page blocks as pagearrays:

```
    def getAsyncJobResult(jobId):
        pages = []
        response = textract.get_document_text_
    detection(JobId=jobId)
        pages.append(response)
        ntoken = None
        if('NextToken' in response):
```

```
        ntoken = response['NextToken']
    while(ntoken):
        response = textract.get_document_text_
detection(JobId=jobId, NextToken=ntoken)
        pages.append(response)
        print("Resultset page recieved: {}".
format(len(pages)))
        nextToken = None
        if('NextToken' in response):
            ntoken = response['NextToken']
    return pages
```

Let's see how we can call into previously mentioned process documents asynchronously:

1. We first call into the `startasynJob()` function to start the asynchronous extraction of our document. We pass in our bucket name and the document name.

2. We check for the completion status of the Textract asynchronous job.

3. Once the status is complete, we get the `response` object from the API and loop through the JSON response to print all the line:

```
jobId = startasyncJob(s3BucketName, chapter2_
syncdensedoc)
print("Started job with id: {}".format(jobId))
if(isAsyncJobComplete(jobId)):
    response = getAsyncJobResult(jobId)
# Print detected text
for resultPage in response:
    for item in resultPage["Blocks"]:
        if item["BlockType"] == "LINE":
            print ('\033[94m' + item["Text"] +
'\033[0m')
```

You can check the output here:

```
Started job with id: 81b88a4fd04693a7ac9902454de8b9697ae70d950f41233cac56da207b8a058b
Job status: IN_PROGRESS
Job status: IN_PROGRESS
Job status: SUCCEEDED
It was a Wednesday evening, and I was collecting all my receipts and busy filling out my
insurance claim document. I wanted to submit to my Health Insurance for reimbursement for
my
COVID-19
test kits that I have purchased. The next day I went to the post office to send
documents through postal mail to my insurance provider. This made me think, in the 21° st
century we are still working with physical documents. With my approximate math this month
alone, we will get X number of documents per month considering 20% of entire US
population buys a test kit. This is a tons of documents as in this instance. In addition to
physical
copies we have tons of documents which are might be just scanned documents. And
we
are looking for manual processing for these number of documents. Can we do any better
in 21st century to automate the process of these documents?
```

Figure 2.4 – LINES output of the asynchronous API

Let's move on to sensitive document processing next.

Sensitive document processing

At times, some companies have their own internal security and regulations and do not want to store documents in the cloud. Other times, a customer already has a data store on-premises and does not want to store documents in two different data stores. That adds work to maintain two different data stores, one in the cloud and the other on-premises. Can they process documents without sending them to the cloud?

So, let's check the solution in code:

1. We are using a local file, `sensitive-doc.png`, for our testing. We are not uploading the document to Amazon S3, but we want to process it locally by calling into the Amazon S3 API:

    ```
    # Document Name
    chapter2_sensitivedoc = "sensitive-doc.png"
    print(chapter2_sensitivedoc)
    ```

2. Let's first check what our `sensitive-doc.png` looks like:

    ```
    display(Image(filename=chapter2_sensitivedoc))
    ```

Let's check the following output:

This is really sensitive information

I store this file in my local file system and storing in another data store may require additional maintenance for my team.

So I want to directly process this document without storing in cloud

Figure 2.5 – Sensitive Document content display

3. Next, we collect all the contents from the documents as raw bytes.

4. We pass these raw bytes as a parameter to Amazon Textract's synchronous API-detect_ document_text API.

5. After processing, we get the response from the synchronous API in the response object.

6. We loop through the JSON response and collect all LINE items.

7. Finally, we print all the LINE items from the sensitive-doc.png document:

```
# Read sensitive document content
with open(chapter2_sensitivedoc, 'rb') as document:
    imgbytes = bytearray(document.read())
# Call Amazon Textract by sending bytes
res = textract.detect_document_text(Document={'Bytes':
imgbytes})
# Print detected text
for line in res["Blocks"]:
    if line["BlockType"] == "LINE":
        print (line["Text"])
```

Let's check the following output:

```
This is really sensitive information.
I store this file in my local filesystem and storing it in another data store may require
additional maintenance for my team.
So I want to directly process this document without storing it in the cloud.
```

Figure 2.6 – The lines output from the sensitive document

Document diversity

Documents come in different formats, layouts, types, and sizes. Some documents can just be a single page type, but others can be of multiple pages and can be of type PDF, JPEG, PNG, TIFF, Word, and so on. Examples of single-page documents include driver's licenses and insurance cards, and examples of multipage documents include legal contract documents and policy documents.

These documents can have diverse layouts, such as a structured layout (such as a table) or unstructured (such as dense paragraphs of text). Alternatively, they can be semi-structured, such as form type of layouts in documents. Sometimes, we get documents that can have multiple different formats, such as dense text interweaved with forms and tables in single or multiple pages. We will look into the accurate extraction of these different formats in *Chapter 3, Accurate Document Extraction with Amazon Textract.*

Preprocessing for document processing

Our ML models may require a certain quality and prerequisite before the accurate extraction of data elements from documents. Some of the prerequisites might include a minimum font size requirement or a maximum document size for accurate document processing. For example, Amazon Textract can process documents with a maximum size of 10 MB and a recommended resolution of 150 dpi. You can find all the quotas and limits at the following service limit page link: `https://docs.aws.amazon.com/textract/latest/dg/limits.html`.

For those instances, you will have to take additional steps to preprocess those documents. We will not be covering preprocessing for IDP in detail, but we highly recommend you to check out the blog post here: `https://aws.amazon.com/blogs/machine-learning/improve-newspaper-digitalization-efficacy-with-a-generic-document-segmentation-tool-using-amazon-textract/`.

Understanding document classification with the Amazon Comprehend custom classifier

Sometimes, we receive many documents as a single package, and we need to process each document individually to derive insights from it as per business requirements. To achieve this, one of the major tasks is to categorize and index different types of documents. This later helps in the accurate extraction of information that meets business-specific requirements. This process of categorizing documents is known as **document classification**.

Let's look into a claims processing use case in the insurance industry. Claims processing is very much a transactional use case, with millions of claims getting processed every single year in the United States. A manual submission of a claims package is a combination of multiple documents, such as an insurance form, receipts, invoices, ID documents, and some unstructured documents such as doctor's notes and discharge summaries. A package is a combination of scanned pages, where each document can be a single or multiple pages long. You want to categorize these documents accurately for the accurate extraction of information. For example, for an invoice document, you may want to extract the amount and invoice ID, but for an insurance form, you may want to extract a diagnosis code and a person's insurance ID number. So, the extraction requirements depend on the type of document you are processing. For that, we need to classify documents before processing. With a manual workforce, you can preview the documents to categorize them into their specified folders. But as with any other manual processing, it can be expensive, error-prone, and time-consuming, so we need an automated way to classify these documents. The following diagram shows a process to classify documents.

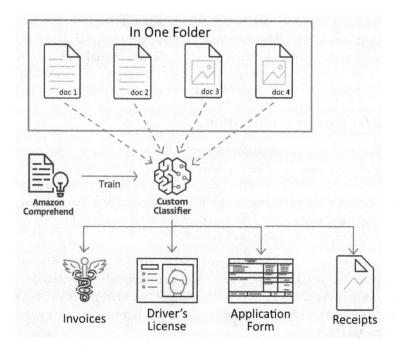

Figure 2.7 – The IDP pipeline – data classification

This process is crucial when we try to automate our document extraction process, where we receive multiple documents and don't have a clear way to identify each document type. If you are dealing with a single document type or have an identifiable way to locate a document, then you can skip this step in the IDP pipeline. Alternatively, classify those documents correctly before proceeding in the IDP pipeline.

We will use the **Amazon Comprehend custom classifier** feature for automating our document classification process. Amazon Comprehend provides a capability to bring in your labeled documents to leverage Comprehend's AutoML feature to accurately classify documents. The auto-algorithm selection and model tuning significantly helps you spend less time building your classification model. Moreover, the ML infrastructure is already taken care of by Amazon Comprehend. As it is a managed service, it automatically takes care of your ML infrastructure to build, train, and deploy your model. You do not need to spend time and effort building your ML platform to train your model. After Amazon Comprehend builds a classification model from your dataset, it gives you the evaluation metrics for your model performance.

Custom classification is a two-step process:

1. You train your own custom classification model to identify classes as per your requirement.
2. You send unlabeled documents for accurate classification.

In this architecture, we will walk you through the following steps:

1. The user uploads documents to be classified.

2. Amazon Textract extracts data from each document. We will discuss the Amazon Textract extraction feature in detail in *Chapter 3, Accurate Document Extraction with Amazon Textract*.

3. Extracted data is sent to the Amazon Comprehend real-time classifier to accurately categorize documents.

Let's now dive into the details of training a Comprehend custom classifier.

Training a Comprehend custom classification model

First, we will walk you through the following architecture on how you can train a custom classification model and run an analysis job for inferencing or classifying documents.

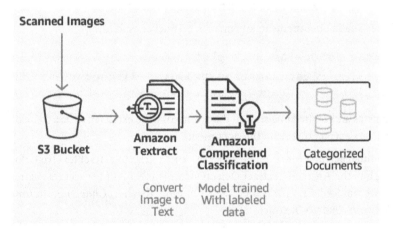

Figure 2.8 – The IDP pipeline – data classification architecture

This architecture walks you through the following steps:

1. Training documents such as insurance forms, invoices, and ID documents are already uploaded into an Amazon S3 bucket during the data capture stage, explained previously.

2. Amazon Textract extracts the text from these documents. We will discuss the Amazon Textract extraction feature in detail in *Chapter 4, Accurate Extraction with Amazon Comprehend*.

3. We perform additional post-processing to create labeled training data for the classification training model.

4. Using this labeled training data, an Amazon Comprehend job is created to classify documents.

5. After training is completed, we create an analysis job to classify documents as required by our categorization logic.

> **Important note**
> Amazon Comprehend also offers a real-time endpoint mode for inference of documents. Comprehend charges per resource endpoint, which can be expensive, so it is recommended to delete your endpoint when not in use. Also, if your workload can be processed asynchronously, I recommend using a Comprehend analysis job instead of creating a real-time endpoint.

Now, let's look at a hands-on example. For the dataset, we will be leveraging invoice- and receipt-type documents. Also, for full notebook walkthrough check: `https://github.com/PacktPublishing/Intelligent-Document-Processing-with-AWS-AI-ML-/blob/main/chapter-2/Chapter-2-comprehend-classification.ipynb`.

1. Let's start with the dataset. All training datasets can be found in the `classification-training` folder on GitHub: `https://github.com/PacktPublishing/Intelligent-Document-Processing-with-AWS-AI-ML-/tree/main/chapter-2/classification-training-dataset`

2. Upload the sample documents to an Amazon S3 bucket as follows:

    ```
    # Upload images to S3 bucket:
    !aws s3 cp classification-training-dataset s3://{data_
    bucket}/idp/textract --recursive --only-show-errors
    ```

    The documents are scanned images, and the text needs to be extracted to be a trained classification model. So, let's extract text from the documents.

3. We have written a helper function, `textract_extract`, to extract text from documents. We are calling into the Amazon Textract DetectDocumentText API for accurate extraction. Please note that we will look into Amazon Textract in more detail in *Chapter 3, Accurate Document Extraction with Amazon Textract*:

    ```
    def textract_extract(table, bucket=data_bucket):
        try:
            response = textract.detect_document_text(
                Document={
                    'S3Object': {
                        'Bucket': bucket,
                        'Name': table
                    }
                })
    ```

4. We process the extracted text and stored it in the right folders for model labeling.

5. We store invoice sample documents in the `invoices` folder and the receipt sample documents in the `receipts` folder. Then, we call the helper function, `data_path`, which loops through the directory and creates a dictionary to hold the extracted document path.

The sample output is shown as follows:

25	receipt-training	THE AIML StORE 1234 SOMEWHERE RD POWAY, CALIFO...
26	receipt-training	THE AIML StORE 1234 SOMEWHERE RD POWAY, CALIFO...
27	receipt-training	THE AIML StORE 1234 SOMEWHERE RD POWAY, CALIFO...
28	receipt-training	THE AIML StORE 1234 SOMEWHERE RD POWAY, CALIFO...
29	receipt-training	THE AIML StORE 1234 SOMEWHERE RD POWAY, CALIFO...
30	invoice-dataset	Invoice #: 96835747-6 INVOICE Created: Dec 09,...
31	invoice-dataset	Invoice #: 24315125-2 INVOICE Created: Jul 30,...

Figure 2.9 – Labeled sample documents

6. After the documents are labeled, we upload the training dataset to an Amazon S3 bucket, as shown here:

```
# Upload Comprehend training data to S3
key='idp/comprehend/comprehend_train_data.csv'
data_compre.to_csv("comprehend_train_data.csv",
index=False, header=False)
s3.upload_file(Filename='comprehend_train_data.csv',
                Bucket=data_bucket,
                Key=key)
```

7. Then, we call into the Amazon Comprehend create_document_classifier API to train the classification model:

```
create_response = comprehend.create_document_
classifier(
        InputDataConfig={
            'DataFormat': 'COMPREHEND_CSV',
            'S3Uri': f's3://{data_bucket}/{key}'
        },
        DataAccessRoleArn=role,
        DocumentClassifierName=document_classifier_name,
        VersionName=document_classifier_version,
        LanguageCode='en',
        Mode='MULTI_CLASS'
    )
document_classifier_arn = create_
```

```
response['DocumentClassifierArn']
    print(f"Comprehend Custom Classifier created with
ARN: {document_classifier_arn}")
```

8. Comprehend can take time to train your model. You can use Amazon Comprehend's describe_document_classifier() command or check on the AWS Management Console for the completion status.

After the Comprehend classifier model is trained, you can check the performance of the **Comprehend classification model** as follows:

1. Go to **Custom classification** on the AWS Management Console of Amazon Comprehend.

Figure 2.10 – Amazon Comprehend – Custom classification

2. Select the Comprehend Custom classification model that we trained in the previous steps.

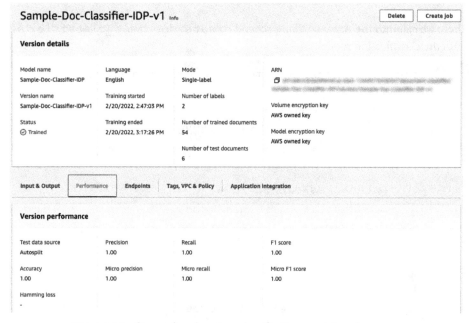

Figure 2.11 – Comprehend custom classification model performance

It is recommended to check the Comprehend training performance metrics at the following link: https://docs.aws.amazon.com/comprehend/latest/dg/cer-doc-class.html.

Let's next move on to cover inference with the Comprehend classifier.

Inference with the Comprehend classifier

We will submit a sample document to our trained custom classifier and leverage a Comprehend analysis job for inference.

Amazon Comprehend requires a text document with UTF-8 encoding. We have already processed the sample document and stored it in https://github.com/PacktPublishing/Intelligent-Document-Processing-with-AWS-AI-ML-/blob/main/chapter-2/rawText-invoice-for-inference.txt.

If you want to use your document, which can be of any scanned image type, we recommend calling Amazon Textract to extract text from it, as follows:

```
response = textract.detect_document_text(
    Document={
        'S3Object': {
            'Bucket': bucket,
            'Name': table
        }
    })
```

After data is extracted from the document by Amazon Textract, go to Amazon Comprehend in the AWS Management Console and follow these steps:

1. Click on **Create Analysis job**.

2. In **Job settings**, select the classification model that you trained in the previous steps.

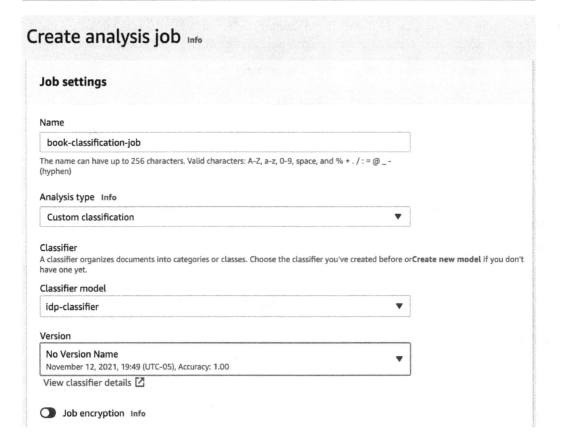

Figure 2.12 – Creating an Amazon Comprehend classification job

3. Provide the input data path of the aforementioned sample document.

4. Select your S3 bucket for output data; this is where the result will be stored.

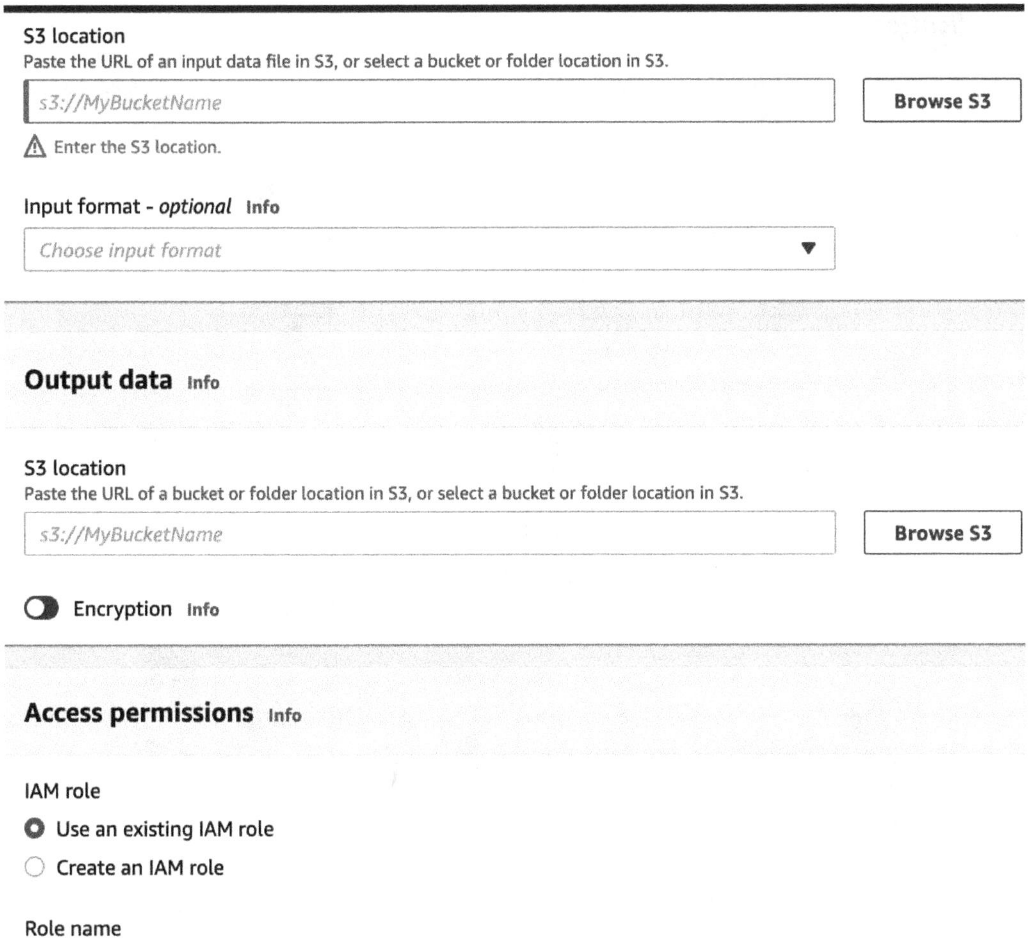

S3 location

Paste the URL of an input data file in S3, or select a bucket or folder location in S3.

| s3://MyBucketName | **Browse S3** |

⚠ Enter the S3 location.

Input format - *optional* Info

| Choose input format ▼ |

Output data Info

S3 location

Paste the URL of a bucket or folder location in S3, or select a bucket or folder location in S3.

| s3://MyBucketName | **Browse S3** |

⬤ Encryption Info

Access permissions Info

IAM role

🔘 Use an existing IAM role

◯ Create an IAM role

Role name

Figure 2.13 – Filling in classification job information

5. Select **Create new IAM role**.

6. Then, click **Create job**.

Let's look at the result next.

Result

After the analysis job is completed, go to the Amazon S3 bucket that you have specified as the output location. Check the result in the output.tgz file.

Classification of documents based on sensitive data

One of the common use cases across regulatory industries is classifying documents based on the presence of sensitive data in them.

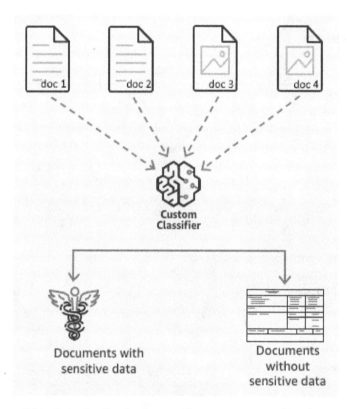

Figure 2.14 – Data classification for sensitive versus non-sensitive documents

We will dive into a hands-on solution in *Chapter 4, Accurate Extraction with Amazon Comprehend*, in the *How to leverage document enrichment for document classification* section.

Understanding document categorization with computer vision

In document preprocessing, you will come across use cases where documents are classified based on a branded logo on the document or having tables. These are visual clues that we want to use to classify documents before processing. We can use **Amazon Rekognition**, computer vision software with deep learning-powered image recognition, to detect visual clues such as objects, scenes, and text from any scanned images for document classification.

You can leverage an **Amazon Rekognition Custom Label** to detect a logo from any document and classify the document based on the logo. For example, a healthcare provider supports multiple insurance providers. Patients when visiting a doctor's office submit insurance cards. These insurance cards can be processed automatically to detect the logos from them, and the documents can be classified according to their corresponding categories.

Now, let's look at a hands-on example, where we have input documents with an AWS logo as well as a non-AWS logo:

1. We will train a Rekognition model to identify the AWS logo from a document.

2. After the model is trained, we will leverage the trained model to classify documents into two categories, AWS documents and non-AWS documents.

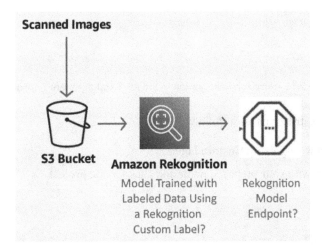

Figure 2.15 – Data classification of structural elements

3. Go to **Amazon Rekognition** in the AWS Management Console. We will be covering the following steps, as shown in the following figure, to train a Rekognition model.

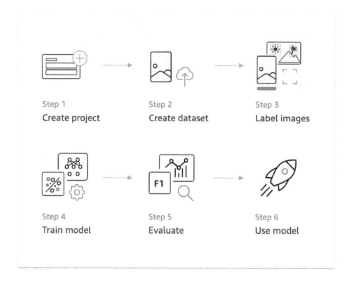

Figure 2.16 – Amazon Rekognition in the AWS Management Console

The data preparation instructions are as follows:

1. Go to **Amazon Rekognition Custom Label.**

2. Create a project with your preferred name and click **Create project**.

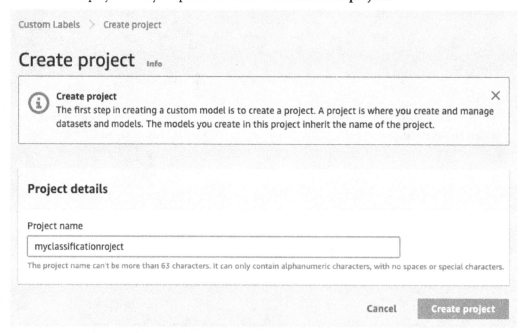

Figure 2.17 – An Amazon Rekognition project

3. Next, click on **Create dataset**. You can upload these <git link> documents as an initial dataset.

> **Important note**
> We recommend using a more versatile and equal distribution of samples for accurate training of the model.

4. Next, add labels to your images.

5. Click **Start Labeling Job** and follow the instructions to label your images.

    For labeling, we are using the Amazon Rekognition console to tag the logo appropriately, as shown here:

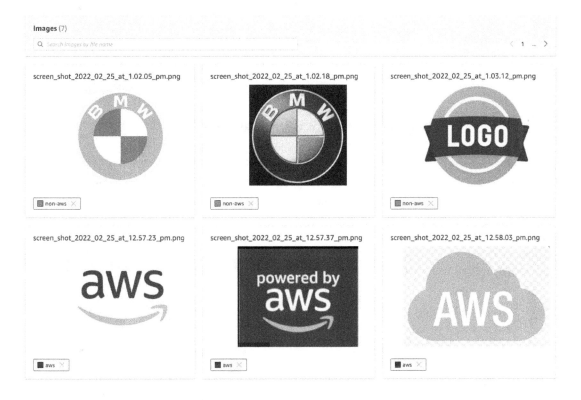

Figure 2.18 – Amazon Rekognition labeling

Next, follow these Rekognition model training instructions:

1. Once the labeling task is completed, click on **Train model**.

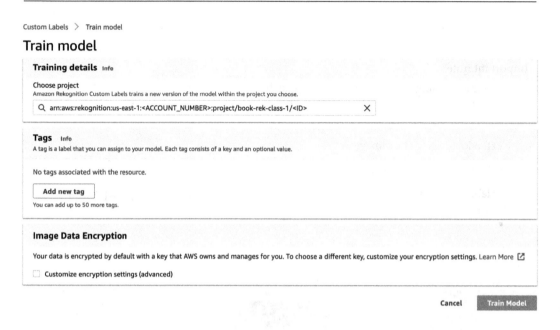

Figure 2.19 – Amazon Rekognition model training

It will take some time to train your Rekognition model.

2. Rekognition, with the default setting, will split the dataset into train and testing datasets.

3. Click on the **Use model** tab shown in *Figure 2.20* and start your model. Wait for the model to come to the *RUNNING* state.

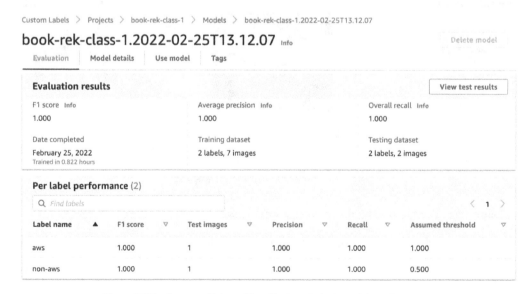

Figure 2.20 – Amazon Rekognition Custom Label model performance

4. Now, let's get inference from our trained model. You can find full walk-through of the Notebook here: `https://github.com/PacktPublishing/Intelligent-Document-Processing-with-AWS-AI-ML-/blob/main/chapter-2/Chapter-2-Rekognition-Classification.ipynb`. Go through the following code block to import the required libraries:

```
import boto3
import io
import logging
import argparse
from PIL import Image, ImageDraw, ImageFont
from botocore.exceptions import ClientError
```

5. Here is the method to call the Rekognition Custom Label model:

```
def callrek_cl(document):
    model="arn:aws:rekognition:us-east-1:<MY_
ACCOUNT>:project/book-rek-class-1/version/book-rek-class-
1.2022-02-25T13.12.07/1645812727191"
    image=Image.open(document)
    image_type=Image.MIME[image.format]
    # get images bytes for call to classify
    image_bytes = io.BytesIO()
    image.save(image_bytes, format=image.format)
    image_bytes = image_bytes.getvalue()
    response = rek_client.detect_custom_
labels(Image={'Bytes': image_bytes},
ProjectVersionArn=model)
    return response
```

6. Call the method defined previously:

```
response = callrek_cl("chapter-2-rekcl.png")
print(response)
response1 = callrek_cl("cha15train.png")
print(response1)
```

Check the output response shown here.

{'CustomLabels': [{'Name': 'aws', 'Confidence': 94.83900451660156}], 'ResponseMetadata': {'RequestId': '9384532f-ca
7b-44a8-ae04-dc8d11a4d6e2', 'HTTPStatusCode': 200, 'HTTPHeaders': {'x-amzn-requestid': '9384532f-ca7b-44a8-ae04-dc8
d11a4d6e2', 'content-type': 'application/x-amz-json-1.1', 'content-length': '64', 'date': 'Fri, 04 Mar 2022 21:09:5
8 GMT'}, 'RetryAttempts': 0}}
{'CustomLabels': [{'Name': 'non-aws', 'Confidence': 38.64500045776367}], 'ResponseMetadata': {'RequestId': 'a030fe1
b-9e92-44c6-9dee-0a66fe5ff7ea', 'HTTPStatusCode': 200, 'HTTPHeaders': {'x-amzn-requestid': 'a030fe1b-9e92-44c6-9dee
-0a66fe5ff7ea', 'content-type': 'application/x-amz-json-1.1', 'content-length': '68', 'date': 'Fri, 04 Mar 2022 21:
09:58 GMT'}, 'RetryAttempts': 0}}

Figure 2.21 – Amazon Rekognition classification output

Summary

In this chapter, we discussed how to build a data capture stage for the IDP pipeline. Data is your gold mine, and you need a secure, scalable, and reliable data store. We introduced Amazon S3 and how you can leverage an object store to aggregate and store data in a scalable and highly available manner. We also described the data capture stage, with documents of varying layouts, formats, and types.

We then reviewed the need for document classification and categorization, with examples including mortgage processing and insurance claims processing. We discussed Amazon Comprehend and its custom classification feature. This chapter also gave you a hands-on experience in how to classify documents as invoice and receipt types. Moreover, we also looked at Amazon Rekognition, and how we can use its Custom Label feature to classify documents on its structural formats. You also had hands-on experience in classifying documents with the presence of the AWS logo or a non-AWS logo.

In the next chapter, we will go through the details of document extraction. We will look into the details of the extraction stage in the IDP pipeline. We will also dive deep into one of the AI services, Amazon Textract, to accurately extract information from any type of document.

Accurate Document Extraction with Amazon Textract

In the previous chapter, you read how businesses can use **Amazon Simple Storage Service (Amazon S3)** and the **Amazon Comprehend custom classifier** to collect and categorize documents before accurate document extraction. We will now dive into the details of the extraction stage of the **Intelligent Document Processing (IDP)** pipeline for the accurate extraction of documents. We will navigate through the following sections in this chapter:

- Understanding the challenges in legacy document extraction
- Using Amazon Textract for accurate extraction of different types of documents
- Using Amazon Textract for the accurate extraction of specialized documents

Technical requirements

For this chapter, you will need access to an **Amazon Web Services (AWS)** account. Before getting started, we recommend that you create an AWS account by referring to the AWS account setup and Jupyter notebook creation steps, as mentioned in the *Technical requirements* section of *Chapter 2, Document Capture and Categorization*. You can find Chapter 3 code samples in GitHub here: `https://github.com/PacktPublishing/Intelligent-Document-Processing-with-AWS-AI-ML-/tree/main/chapter-3`

Understanding the challenges in legacy document extraction

Many organizations, across industries, irrespective of the size of the business, deal with a large number of documents in everyday transactions. Moreover, we discussed the data diversity, data sources, and various layouts and formats for these documents in *Chapter 2, Document Capture and Categorization*.

The data diversity at scale makes it difficult to extract elements from these documents. For example, think about a back-office task for a company. This is one of the non-mission critical tasks for a company, but at the same time, these tasks need to be fulfilled in a scheduled and timely manner. For example, the back office receives invoices at scale and needs to extract information and put it in a structured way in its **enterprise resource planning** (**ERP**) system, such as **Systems, Applications, and Products** (**SAP**) for accurate payments. Once we convert the data from unstructured documents to a structured format, a machine can handle the processing. So, without automation, how do we handle the extraction part of document processing?

One of the most common ways to extract documents is through **manual processing**.

Human beings can manually scan these documents and extract elements and feed them to their data entry system, but manual processing of documents can be error-prone, time-consuming, and expensive. Alternatively, customers can use traditional technologies such as **optical character recognition** (**OCR**) to extract all elements from the documents. But OCR can give you a flat **bag of words** (**BoW**), which is not an optimal way to extract documents. Additionally, we have seen customers using template- or rule-based document extraction. But template-based extraction can be high maintenance, and to onboard a new document type or new format, you will have to create and manage its corresponding template. Moreover, if you have used your custom model building process to extract elements from your documents, to onboard a new document type, it takes significant time and effort to retrain the technology stack for accurate extraction. Moreover, most of the time, traditional document processing—including manual processing—is not ready to handle document extraction at scale. Is there any other alternative solution for accurate document extraction? The following diagram depicts a sample architecture for document extraction in an IDP pipeline:

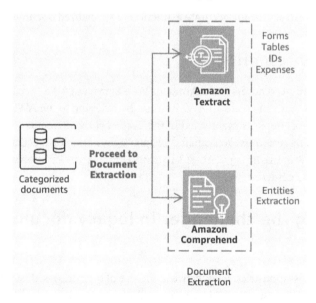

Figure 3.1 – Document extraction in IDP pipeline

In the document extraction stage of the IDP pipeline, our goal is to extract all the fields accurately from any type of document. In *Figure 3.1*, after categorizing the documents, we want to extract elements unique to that document. For example, you might want to get the **Social Security number** (**SSN**) from a **Social Security card** (**SSC**), but the Bank account number from a bank statement document. We may use different technologies and AWS **artificial intelligence** (**AI**) services such as Amazon Textract and Amazon Comprehend for accurate extraction. Before diving deep into the architecture, let's first learn how we can use Amazon Textract for accurate document extraction.

Using Amazon Textract for the accurate extraction of different types of documents

To build an IDP pipeline, we need an AI service such as Amazon Textract to accurately extract printed as well as handwritten text from any type of document. Now, let's look into the features of Amazon Textract for accurate document extraction.

Introducing Amazon Textract

Amazon Textract is a fully managed AI/**machine learning** (**ML**) service, built with pre-trained models for accurate document extraction. Amazon Textract goes beyond simple OCR and can extract data from any type of document accurately. To leverage Textract, no ML experience is required; you can call into its **application programming interfaces** (**APIs**) to leverage the pre-trained ML model behind the scenes.

Amazon Textract provides APIs that you can call through the serverless environment without the need to manage any kind of infrastructure. Once the document content is extracted, you can proceed to the subsequent stages of the IDP pipeline, as required by your business. As with any other ML model, the Textract model learns from the data and gives intelligent results over time, but the main benefit of leveraging a managed service such as Textract is that, with no infrastructure to manage, it can still get accurate results just by calling into its API. Now, let's see some of the common benefits of accessing managed models through serverless APIs in the following table:

	Build your own ML model (for IDP pipeline)	ML model that you can customize (for IDP pipeline—for example, Comprehend custom classifier and custom entity recognizer)	Pre-trained ML model (for IDP pipeline—for example, Amazon Textract, Amazon Comprehend pre-trained, Amazon Comprehend Medical)
ML skillset	Required	Basic understanding required	Not required
Data preparation	Required	Required but minimal samples required	Limited requirement Most of the time, data preprocessing is not required, but sometimes it is required, as described in *Chapter 2, Document Capture and Categorization*, in the *Preprocessing for document processing* section
Model selection, training, and tuning	Required	Limited requirement	Not required
Model deployment	Required	Not required	Not required
Model monitoring; maintenance	Required	Not required for the infrastructure but for data drift, you may need to retrain	Not required
Model inference infrastructure	Required	Not required	Not required
Time to market (TTM)	Longer	Relatively less with **automated ML (AutoML)** capabilities	Fastest
Cost	Cost of infrastructure, and data science skillset	Pay as you go for inference	Pay as you go

Table 3.2 – ML model versus pre-trained AI services

The preceding table gives you a high-level comparison of the benefits of using a managed ML model for your document extraction. As a business decision maker, you can map it back to your timeline requirement and priority as per business requirements. This will help you define how much ML investment you will do for your team. If you decide to go with managed ML model(s), you can save time and can deliver results to the market faster. You can then leverage the saved time for your additional priorities.

Accurate extraction of unstructured document types

Unstructured documents are dense text types, such as legal contractual documents. For the hands-on lab for this chapter, we will be using the following sample text. A snippet of the dense text document is provided here:

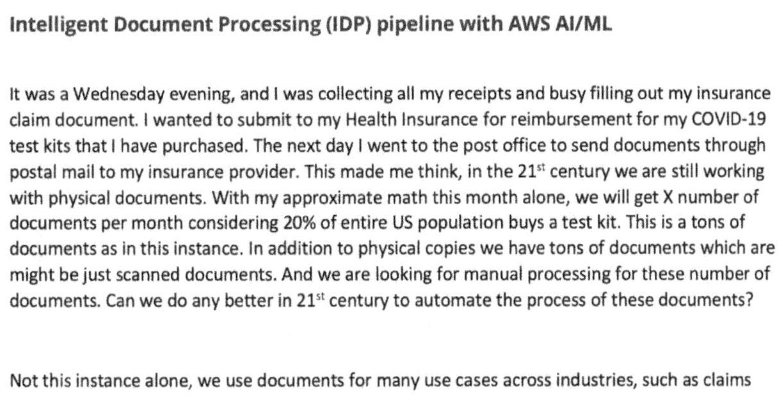

Intelligent Document Processing (IDP) pipeline with AWS AI/ML

It was a Wednesday evening, and I was collecting all my receipts and busy filling out my insurance claim document. I wanted to submit to my Health Insurance for reimbursement for my COVID-19 test kits that I have purchased. The next day I went to the post office to send documents through postal mail to my insurance provider. This made me think, in the 21st century we are still working with physical documents. With my approximate math this month alone, we will get X number of documents per month considering 20% of entire US population buys a test kit. This is a tons of documents as in this instance. In addition to physical copies we have tons of documents which are might be just scanned documents. And we are looking for manual processing for these number of documents. Can we do any better in 21st century to automate the process of these documents?

Not this instance alone, we use documents for many use cases across industries, such as claims

Figure 3.3 – Dense text document example

We will follow the next steps to accurately extract elements using Amazon Textract. You can follow the Notebook for your Document Extraction: `https://github.com/PacktPublishing/Intelligent-Document-Processing-with-AWS-AI-ML-/blob/main/chapter-3/Textract-chapter3.ipynb`

1. Firstly, we will start out by importing the required libraries, such as `boto3`, as follows:

```
import boto3
import os
from io import BytesIO
from PIL import Image
from IPython.display import Image, display, JSON, IFrame
from trp import Document
from PIL import Image as PImage, ImageDraw
from IPython.display import IFrame
```

2. Let's see what our `unstructured.png` sample document looks like. You can see this here:

```
# Document
documentName = "unstructured.png"
display(Image(filename=documentName))
```

3. Get the Textract `boto3` client by executing the following code:

```
#Extract dense Text from scanned document
# Amazon Textract client
textract = boto3.client('textract')
```

4. To extract from dense text types of documents, we are using the `detect_document_text` API of Amazon Textract, as shown in the following code snippet. This Amazon Textract API can accept a local file or an S3 object as the input document. For our code here, we are using a local file as an example. Thus, we are passing raw bytes from the local file to Textract's API:

```
# Read document content
with open(documentName, 'rb') as document:
    imageBytes = bytearray(document.read())
# Call Amazon Textract
response = textract.detect_document_
text(Document={'Bytes': imageBytes})
```

5. Now, let's print Textract's response as `LINES`. The code and sample output are shown in the following snippet:

```
# Print detected text
for item in response["Blocks"]:
    if item["BlockType"] == "LINE":
        print (item["Text"])
print(response)
```

Now, let's check the result here:

```
Intelligent Document Processing (IDP) pipeline with AWS AI/ML
It was a Wednesday evening, and I was collecting all my receipts and busy filling out my insurance
claim
document. I wanted to submit to my Health Insurance for reimbursement for my COVID-19
test kits that I have purchased. The next day I went to the post office to send documents through
postal mail to my insurance provider. This made me think, in the 21st century we are still working
with physical documents. With my approximate math this month alone, we will get X number of
documents per month considering 20% of entire US population buys a test kit. This is a tons of
documents as in this instance. In addition to physical copies we have tons of documents which are
might be just scanned documents. And we are looking for manual processing for these number of
documents. Can we do any better in 21st century to automate the process of these
documents?
Not this instance alone, we use documents for many use cases across industries, such as claims
```

Figure 3.4 – Lines from document

The preceding code walked you through how you will call Textract's API directly using Python code for a dense text type of document extraction. Let's now check another way to extract text from documents, leveraging the `amazon textract textractor` library.

The Textractor library helps speed up the code implementation with its inbuilt abstraction logic and response parser. This is available as the `pypi` library.

Prerequisites to use the Textractor `pypi` library include the following:

- Python 3

- AWS **command-line interface (CLI)**

Now, let's check in the next code samples how we can use the Textractor library to extract accurately from the same dense text type of documents. Proceed as follows:

1. Import the required `pypi` library, like so:

```
!python -m pip install amazon-textract-caller
!python -m pip install amazon-textract-prettyprinter
import json
from trp import Document
from textractcaller import call_textract, Textract_
Features
from textractprettyprinter.t_pretty_print import Pretty_
Print_Table_Format, Textract_Pretty_Print, get_string
```

2. Use the `call_textract` method to extract elements from your document. This method takes the local filename directly as input. It abstracts the creation of "raw bytes" from the local document. The user doesn't have to get raw bytes and pass them to the Textract API anymore. We can directly pass the local filename [documentName] for accurate extraction, as illustrated in the following code snippet:

```
textract_json = call_textract(input_
document=documentName)
```

Now, let's print Textract's response as LINES. For printing, we are using the `get_string()` method of the `prettyprinter` library. `prettyprinter` formats the **JavaScript Object Notation (JSON)** output response of Textract for easy reading. In the following code snippet, we are using `Textract_Pretty_Print.LINES` for our document processing, but additional types such as WORDS and more are supported:

```
print(get_string(textract_json=textract_json,
                output_type=[Textract_Pretty_Print.
LINES]))
```

Sample output is shown here:

```
Intelligent Document Processing (IDP) pipeline with AWS AI/ML
It was a Wednesday evening, and I was collecting all my receipts and busy filling out my insurance
claim
document. I wanted to submit to my Health Insurance for reimbursement for my COVID-19
test kits that I have purchased. The next day I went to the post office to send documents through
postal mail to my insurance provider. This made me think, in the 21st century we are still working
with physical documents. With my approximate math this month alone, we will get X number of
documents per month considering 20% of entire US population buys a test kit. This is a tons of
documents as in this instance. In addition to physical copies we have tons of documents which are
might be just scanned documents. And we are looking for manual processing for these number of
documents. Can we do any better in 21st century to automate the process of these
documents?
Not this instance alone, we use documents for many use cases across industries, such as claims
```

Figure 3.5 – Output of Textract

Now, let's check the JSON API response of Amazon Textract for better parsing.

JSON API response

The document metadata will have reference to the number of pages as **pages** and **blocks**. The blocks are organized in a parent-child relationship. For example, a **LINE (Amazon Textract)** block will have a LINE BlockType, which has a reference to two child blocks of the WORD BlockType: one WORD BlockType for **Amazon** and another WORD BlockType for **Textract**. This relationship is represented in the following diagram:

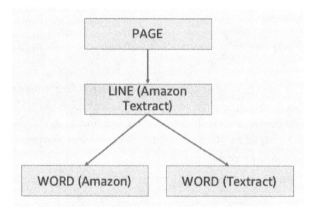

Figure 3.6 – Block relationship for dense text

Now, let's check how we can use Amazon Textract for the extraction of semi-structured/form types of documents.

Accurate extraction of semi-structured document types

Semi-structured documents are of the form type, such as insurance claims forms. They can have key-value elements along with checkbox/radio button types of elements. For the hands-on lab for this chapter, we will be using the following sample document, which shows a snippet of a semi-structured type of document. You can see this document has key-value pairs. Some key-value pairs are on one line, but some values are on multilines below the key. A key value can be separated by a space, and any delimiter also. This document also has a checkbox type:

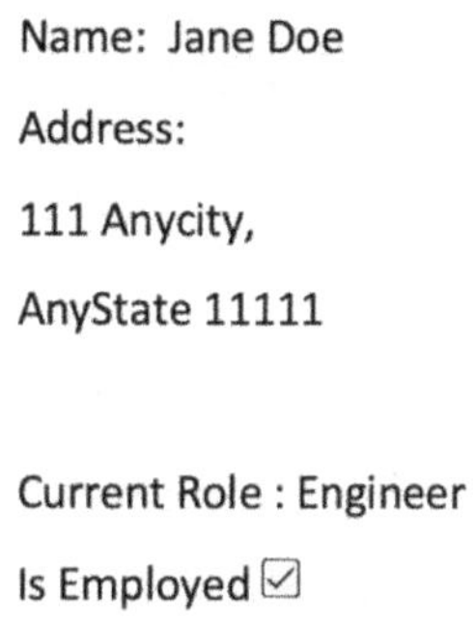

Name: Jane Doe

Address:

111 Anycity,

AnyState 11111

Current Role : Engineer

Is Employed ☑

Figure 3.7 – Semi-structured form-type document

Now, let's look into the following sample code for the accurate extraction of this form type of document. Proceed as follows:

1. Use the following code to display the semi-structured document:

```
# Document
documentName = "semi-structured.png"
display(Image(filename=documentName))
```

This displays the semi-structured sample document, as we can see here:

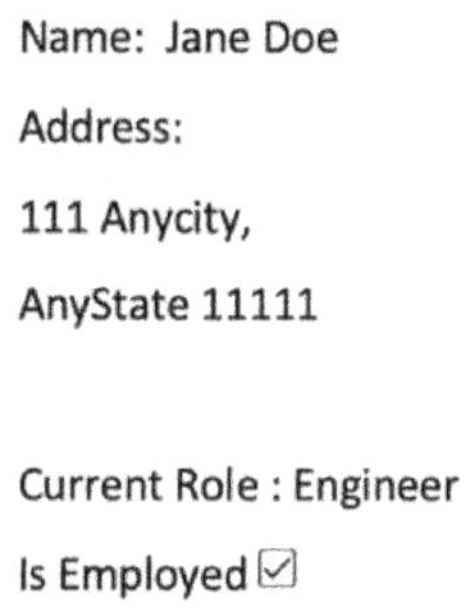

Name: Jane Doe

Address:

111 Anycity,

AnyState 11111

Current Role : Engineer

Is Employed ☑

Figure 3.8 – Sample form-type document

2. Use the `call_textract` method to extract elements from your document. We are passing `Textract_Features` as FORMS to extract from the semi-structured document. This method takes the local filename directly as input. Then, we call the Textractor `prettyprinter` library to print the FORMS type. The code for this is shown here:

```
textract_json = call_textract(input_
document=documentName, features=[Textract_Features.
FORMS])
print(get_string(textract_json=textract_json,
            output_type=[Textract_Pretty_Print.FORMS]))
```

3. In the following output, you can see all the key-value pairs are extracted accurately. Moreover, the checkbox selection is also extracted with a SELECTED value:

```
|-----------------|-----------------------------|
| Key             | Value                       |
| Current Role :  | Engineer                    |
| Is Employed     | SELECTED                    |
| Name:           | Jane Doe                    |
| Address:        | 111 Anycity, AnyState 11111 |
```

Figure 3.9 – Sample Textract output form

Now let's check the JSON API response of Amazon Textract for better parsing.

JSON API response

For a semi-structured document of the form type, if we have a form type, we get a KEY_VALUE_SET BlockType. In the JSON response, we also have EntityType. For a form type, EntityType values can be of KEY or VALUE. Additionally, if we have a checkbox, you will see a BlockType of SELECTION_ELEMENT with SelectionStatus of SELECTED or NOT_SELECTED. Sample output is depicted in the following diagram:

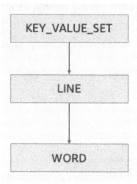

Figure 3.10 – Block relationship for the form type

Now, let's check how we can use Amazon Textract for the extraction of structured/table types of documents.

Accurate extraction of structured document types

Structured documents are table types. They can have a table with rows and columns with cells. Textract table extraction also supports merged cell extraction. For the hands-on lab for this chapter, we will be using the following sample structured type of document. Here, you can see this document has a table with rows and columns:

First Name	Last Name	Job Role	Supervisor Name
Jane	Doe	Engineer	John Doe
John	Doe	Manager	Jane1 Doe1

Figure 3.11 – Sample table type

Proceed as follows:

1. Use the following code for the display of the semi-structured document:

    ```
    # Document
    documentName = "structured.png"
    display (Image(filename=documentName))
    ```

    And the output looks like this:

First Name	Last Name	Job Role	Supervisor Name
Jane	Doe	Engineer	John Doe
John	Doe	Manager	Jane1 Doe1

Figure 3.12 – Sample output: table

2. Use the `call_textract` method to extract elements from your document. We are passing `Textract_Features` as TABLES to extract from the structured document. This method takes the local filename directly as input. Then, we call the Textractor `prettyprinter` library to print the TABLES type. If your document has multiple tables, the following code can extract all the tables from your documents:

    ```
    textract_json = call_textract(input_
    document=documentName, features=[Textract_Features.
    TABLES])
    ```

```
print(get_string(textract_json=textract_json,
                 output_type=[Textract_Pretty_Print.
TABLES]))
```

In the following sample output, you can see a table with headers, rows, and columns extracted from the document:

```
|-------------|------------|-----------|------------------|
| First Name  | Last Name  | Job Role  | Supervisor Name  |
| Jane        | Doe        | Engineer  | John Doe         |
| John        | Doe        | Manager   | Jane1 Doe1       |
```

Figure 3.13 – Sample output: table result

Now, let's check the JSON API response of Amazon Textract for better parsing.

JSON API response

For a structured document of the table type, we get additional TABLE and CELL BlockType values. In the JSON response, we also have EntityType. For a TABLE type, EntityType values can be of COLUMN_HEADER for the column header. Sample output is depicted in the following diagram:

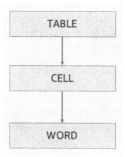

Figure 3.14 – Block relationship for table type

Now, let's check how we can use Amazon Textract for the extraction of specialized documents.

Using Amazon Textract for the accurate extraction of specialized documents

Amazon Textract can support the accurate extraction of specialized document types such as a **United States (US)** driver's license, a US passport, invoices, and receipts. Now, we will walk you through the steps to accurately extract elements from these documents using Amazon Textract.

Accurate extraction of ID document (driver's license)

In this section, we will use Amazon Textract for a US driver's license document extraction. Now, let's dive into the code sample, as follows:

1. We are using `"dl.png"` as a sample US driver's license document. Let's use the following code to display the document:

    ```
    # Document
    documentName = "dl.png"
    display(Image(filename=documentName)
    ```

    Now, check the result, as follows:

Figure 3.15 – Sample driver's license

2. We are using the `analyze_id` synchronous API of Amazon Textract to extract elements from our US driver's license, which is stored locally. To process a local file, we are first extracting the raw bytes from the document and then passing that as an input to the `analyze_id` API for extraction. The code is illustrated in the following snippet:

    ```
    with open(documentName, 'rb') as document:
        imageBytes = bytearray(document.read())
    response = textract.analyze_id(
        DocumentPages=[{"Bytes":imageBytes}]
    )
    print(response)
    ```

3. Use the following code to print the JSON response from the `analyze_id` API:

```
import json
print(json.dumps(response, indent=2))
```

Now, let's check the JSON output, as follows:

```
"IdentityDocuments": [
  {
    "DocumentIndex": 1,
    "IdentityDocumentFields": [
      {
        "Type": {
          "Text": "FIRST_NAME"
        },
        "ValueDetection": {
          "Text": "GARCIA",
          "Confidence": 99.47561645507812
        }
      },
      {
        "Type": {
          "Text": "LAST_NAME"
        },
        "ValueDetection": {
          "Text": "MARIA",
```

Figure 3.16 – Sample JSON output

Amazon Textract can also extract elements from additional **identifier** (**ID**) types of documents such as US passports.

ID document (US passport) accurate extraction

In this section, we will use Amazon Textract for a US passport document extraction. Now, let's dive into the code sample, as follows:

1. We are using `passport.png` as a sample US passport document. Let's use the following code to display the document:

```
# Document
documentName = "passport.png"
display(Image(filename=documentName))
```

Now, check the result, as follows:

Figure 3.17 – Sample passport document

2. We are using the `analyze_id` synchronous API of Amazon Textract to extract elements from our US passport, which is stored locally. To process a local file, we are first extracting the raw bytes from the document and then passing that as an input to the `analyze_id` API for extraction, as illustrated in the following code snippet:

```
with open(documentName, 'rb') as document:
    imageBytes = bytearray(document.read())
response = textract.analyze_id(
    DocumentPages=[{"Bytes":imageBytes}]
)
print(response)
```

3. Use the following code to print the JSON response from the `analyze_id` API:

```
import json
print(json.dumps(response, indent=2))
```

Now, let's check the JSON output, as follows:

```
"IdentityDocuments": [
  {
    "DocumentIndex": 1,
    "IdentityDocumentFields": [
      {
        "Type": {
          "Text": "FIRST_NAME"
        },
        "ValueDetection": {
          "Text": "JUAN",
          "Confidence": 71.55193328857422
        }
      },
      {
        "Type": {
          "Text": "LAST_NAME"
        },
        "ValueDetection": {
          "Text": "JUAN",
```

Figure 3.18 – Sample output

Let's next move on to receipt document accurate extraction.

Receipt document accurate extraction

In this section, we will use Amazon Textract for the accurate extraction of a receipt document. Now, let's dive into the code sample, as follows:

1. We are using `receipt.png` as a sample receipt document. Let's use the following code to display the document:

    ```
    # Document
    documentName = "receipt.png"
    display(Image(filename=documentName))
    ```

Now, check the result, as follows:

THE AIML StORE
1234 SOMEWHERE RD
POWAY, CALIFORNIA 92129
02/13/2022 12:28 AM

9030182301029381-230981

Future Diary: Playing Cards	1	$18.34
Headphones, Sound Intone 165	1	$35.39
Headphones		
Simple Deluxe Clamp Lamp	1	$19.08
Light		

SUBTOTAL	$72.81
T = CA TAX 7.7500 on 78.271	$5.461
TOTAL	$78.271
x4889 AMEX CHARGE	$78.271
AID:	A000088808-4515
	AMEX CREDIT
AUTH CODE:	054515

SOME PROMOTIONS MAY REDUCE THE
REFUND VALUE OF ITEMS

We need your feedback!
Take a quick survey and enter for the chance
To WIN a $250 Store gift card. Go to:
Myfakestore.com/win
5 winners monthly

Figure 3.19 – Sample receipt document

2. We are using the `analyze_expense` synchronous API of Amazon Textract to extract elements from our receipt document, which is stored locally. To process a local file, we are first extracting the raw bytes from the document and then passing that as an input to the `analyze_expense` API for extraction. The code is illustrated in the following snippet:

```
# Read document content
with open(documentName, 'rb') as document:
    imageBytes = bytearray(document.read())
# Call Amazon Textract
response = textract.analyze_expense(Document={'Bytes':
imageBytes})
print(response)
```

3. Use the following code to print the JSON response from the `analyze_expense` API:

```
import json
print(json.dumps(response, indent=2))
```

Now, let's check the JSON output, as follows:

```
"DocumentMetadata": {
  "Pages": 1
},
"ExpenseDocuments": [
  {
    "ExpenseIndex": 1,
    "SummaryFields": [
      {
        "Type": {
          "Text": "VENDOR_NAME",
          "Confidence": 99.84497833251953
        },
        "ValueDetection": {
          "Text": "THE AIML StORE",
          "Geometry": {
            "BoundingBox": {
              "Width": 0.503549337387085,
              "Height": 0.03130369260907173,
              "Left": 0.2559385299682617,
```

Figure 3.20 – Sample output

Invoice document accurate extraction

In this section, we will use Amazon Textract for the accurate extraction and normalization of inferred elements from an invoice type of document. Now, let's dive into the code sample, as follows:

1. We are using `invoice.png` as a sample invoice document. Let's use the following code to display the document:

```
# Document
documentName = "invoice.png"
display(Image(filename=documentName))
```

Now check the result, as follows. I want to quickly highlight that the input document type has skew. This is to show that Amazon Textract can process documents with low resolution as well as documents with skew:

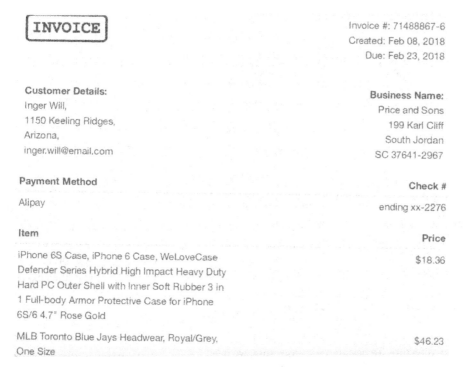

Figure 3.21 – Sample invoice document

2. We are using the `analyze_expense` synchronous API of Amazon Textract to extract elements from our invoice document, which is stored locally. To process a local file, we are first extracting the raw bytes from the document and then passing that as an input to the `analyze_expense` API for extraction. The code is illustrated in the following snippet:

```
# Read document content
with open(documentName, 'rb') as document:
    imageBytes = bytearray(document.read())
# Call Amazon Textract
response = textract.analyze_expense(Document={'Bytes':
imageBytes})
print(response)
```

3. Use the following code to print the JSON response from the `analyze_expense` API:

```
import json
print(json.dumps(response, indent=2))
```

Now, let's check the JSON output, as follows:

```
{'DocumentMetadata': {'Pages': 1}, 'ExpenseDocuments': [{'ExpenseIndex': 1, 'SummaryFields': [{'Type': {'Text': '
OTHER', 'Confidence': 87.5}, 'LabelDetection': {'Text': 'Payment Method', 'Geometry': {'BoundingBox': {'Width':
0.17776434123516083, 'Height': 0.03428881987929344, 'Left': 0.011818407103419304, 'Top': 0.4970138967037201}, 'Po
lygon': [{'X': 0.012267852202057838, 'Y': 0.4970138967037201}, {'X': 0.18958275020122528, 'Y': 0.501367926597595
2}, {'X': 0.18913330137729645, 'Y': 0.5313026905059814}, {'X': 0.011818407103419304, 'Y': 0.5269486904144287}]},
'Confidence': 87.4721908569336}, 'ValueDetection': {'Text': 'Alipay', 'Geometry': {'BoundingBox': {'Width': 0.061
130717396736145, 'Height': 0.030865272507071495, 'Left': 0.010131187736988068, 'Top': 0.5569228529930115}, 'Polyg
on': [{'X': 0.010572228580713272, 'Y': 0.5569228529930115}, {'X': 0.07126190513372421, 'Y': 0.5584130883216858},
{'X': 0.0708208680152893, 'Y': 0.5877881050109863}, {'X': 0.010131187736988068, 'Y': 0.5862978696682312}]}, 'Confi
dence': 87.48966217041016}, 'PageNumber': 1}, {'Type': {'Text': 'OTHER', 'Confidence': 83.0}, 'LabelDetection':
{'Text': 'Check #', 'Geometry': {'BoundingBox': {'Width': 0.08559966832399368, 'Height': 0.02675745263695717, 'Le
```

Figure 3.22 – Sample output

We have been through the extraction stage of IDP and seen how we can leverage Amazon Textract for accurate extraction. Most often, there is a requirement to process millions of documents at scale. Going back to our insurance processing use case for enterprise insurance companies, they receive millions of documents as part of claims processing. Can we process a large number of documents at scale? To answer this, let's look into a large-scale document processing architecture in the next chapter.

Large Scale Document processing

For large-scale document processing, we will use the following Intelligent document processing reference architecture. To design the architecture, we need to take two main things into consideration. One, how many documents you want to process, and second, how fast you want to process them. For this architecture, we want to process tens of thousands of documents per day and our processing is good with not having near real-time processing. For that reason, we are using Amazon Tetxract Asynchronous operation and designing a large-scale architecture.

In the first stage of the IDP (Data Capture) stage, we are storing data in Amazon S3, which kicks in a processor lambda function that can collect a subset of document prefixes. Now you must be thinking how much the subset of document prefix you will define. I would recommend checking Textract **Transaction Per Second** (TPS) limit per your account, region, and **Tetxract API**, and setting the subset of document prefix accordingly not to throttle your Tetxract API. Then in an event-driven manner, store these subsets of documents in Amazon SQS for decoupled architecture. SQS triggers Amazon Tetxract API asynchronous API to process these subsets of documents and notifies the completion status with SNS. Note here, that we are not wasting compute cycle by calling Get Textract Result API in the loop. We are just waiting for Tetxract completion notification, and then call Textract Get result API and store the result back into Amazon S3.

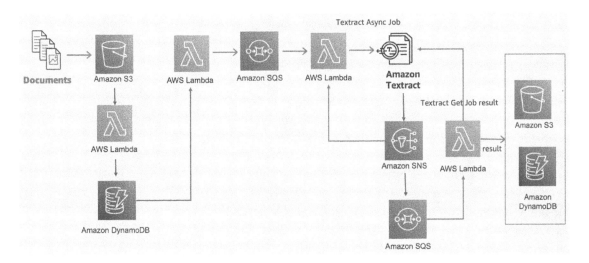

Figure 3.23 – Large Scale Document Processing Architecture

Now that we have looked into a large-scale document processing architecture I would recommend, checking the Textract TPS limit in the following reference: `https://docs.aws.amazon.com/general/latest/gr/textract.html`

Summary

In this chapter, we discussed the extraction stage of an IDP pipeline, and how we can leverage Amazon Textract to accurately extract elements from documents. Documents can be of different types, such as an unstructured dense text type of document, a semi-structured document such as a form, or a structured document such as a table. We walked through the sample code and its API response to accurately extract elements from any type of scanned document.

We then reviewed the need for accurate extraction of elements from specialized document types, such as ID documents such as a US driver's license, a US passport, or invoice/receipt types of documents. We discussed Amazon Textract's `analyze_id` and `analyze_expense` APIs to accurately extract elements from ID and invoice/receipt types of documents respectively. We walked you through the sample code for your accurate extraction of specialized document types.

In the next chapter, we will extend the extraction stage of the document processing pipeline with Amazon Comprehend. Moreover, we will introduce you to the enrichment stage of IDP and how you can leverage Amazon Comprehend to enrich your documents.

4

Accurate Extraction with Amazon Comprehend

In the previous chapter, you learned about the challenges involved in the legacy document extraction process and how businesses can use Amazon Textract for the accurate extraction of elements from any type of document. We will now dive into detailed extraction using Amazon Comprehend for the extraction stage of the **Intelligent Document Processing** (**IDP**) pipeline. We will cover the following in this chapter:

- Using Amazon Comprehend for accurate data extraction
- Understanding document extraction – IDP with Amazon Comprehend
- Understanding custom entities extraction with Amazon Comprehend

Technical requirements

For this chapter, you will need access to an AWS account. Before getting started, we recommend that you create an AWS account by referring to AWS account setup and Jupyter notebook creation steps as mentioned in the Technical requirement section in *chapter 2, Document Capture and Categorization*. You can find Chapter-3 code sample in GitHub: `https://github.com/PacktPublishing/Intelligent-Document-Processing-with-AWS-AI-ML-/tree/main/chapter-4`

Using Amazon Comprehend for accurate data extraction

Amazon Comprehend is a fully managed AWS AI service with pre-trained **Machine Learning** (**ML**) models for accurate data extraction and to derive insights from your documents. It is a managed solution, which means you call Amazon Comprehend's API and pass in your input, and you will get a response back. Amazon Comprehend uses **Natural Language Processing** (**NLP**) to extract meaningful information about the content of unstructured, dense, text content. To use Comprehend, no ML experience is required. You can call its **Application Programming Interfaces** (**APIs**) to leverage the pre-trained ML model behind the scenes. Amazon Comprehend provides APIs that you can call through the serverless environment without the need to manage any kind of infrastructure. Once the document content is extracted, you can proceed with the subsequent stages of the IDP pipeline as required by your business.

Now let's check out the core features and functionalities of Amazon Comprehend.

Amazon Comprehend enables you to examine your unstructured data, for example, unstructured text. It can help you gain various insights about content by using a number of pretrained models.

In *Figure 4.1*, you can see some of the key features of Amazon Comprehend for the document extraction phase of the IDP pipeline.

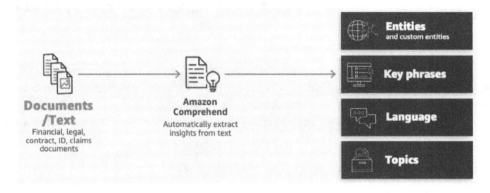

Figure 4.1 – Amazon Comprehend – key features for IDP

Now, let's have a walk-through of these features on the AWS Management Console for Amazon Comprehend:

1. Go to **Amazon Comprehend**. Click on **Launch Amazon Comprehend**:

Figure 4.2 – Amazon Comprehend on the AWS Management Console

2. We will use the following sample text to analyze all of the features of Amazon Comprehend available through the AWS Management Console:

Intelligent Document Processing (IDP) pipeline with AWS AI/ML

It was a Wednesday evening, and Sonali Sahu was collecting all my receipts and busy filling out my insurance claim document. I wanted to submit to my Health Insurance for reimbursement for my COVID-19 test kits that I have purchased. The next day I went to the post office to send documents through postal mail to my insurance provider. This made me think, in the 21st century we are still working with physical documents. With my approximate math this month alone, we will get X number of documents per month considering 20% of entire US population buys a test kit. This is a tons of documents as in this instance. In addition to physical copies we have tons of documents which are might be just scanned documents. And we are looking for manual processing for these number of documents. Can we do any better in 21st century to automate the process of these documents?

Figure 4.3 – Sample unstructured text

3. Copy the preceding text and insert it into **Real-time analysis → Input text**, as shown in *Figure 4.4*, and click on **Built-in** and then **Analyze**:

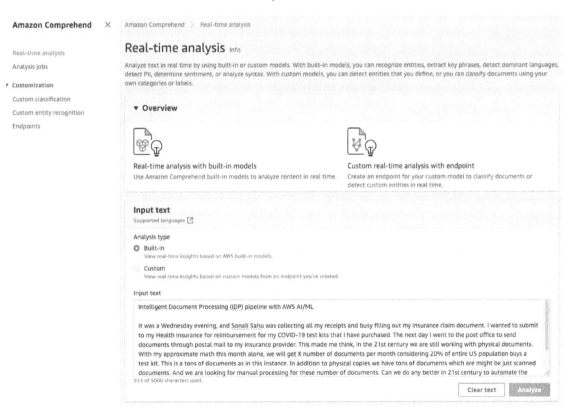

Figure 4.4 – Amazon Comprehend Real-time analysis

4. Scroll down to see the insights.

Now, we will walk through each Insights API by changing each tab:

- **Detecting entities**: In *Chapter 3*, *Accurate Document Extraction with Amazon Textract*, we looked into how all the fields can be extracted from a document. But what if you are just interested in extracting only a few fields? For example, say you have a bank statement, and you might just be interested in extracting and reviewing the date and names from this document. In that case, you can call Amazon Comprehend's pre-trained Detect Entities API to get only person's name and dates from the bank statement document. Now let's enter some sample text as in *Figure 4.3* and call the Amazon Comprehend detect entities feature. You can see from the screenshot in *Figure 4.5* that Amazon Comprehend was able to detect the highlighted entities from the text you entered. Some of the pre-trained entities are **Person**, **Date**, **Location**, **Organization**, and **Date**.

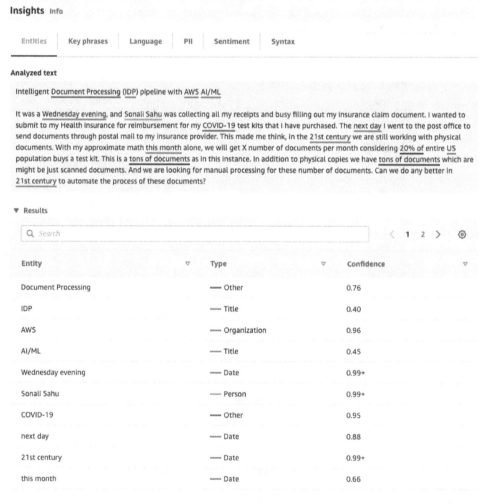

Figure 4.5 – Amazon Comprehend pre-trained detection entities

You can click on subsequent pages on the result screen to see the additional extracted entities. Following is the output from Amazon Comprehend pre-trained Named Entities Recognition additional output.

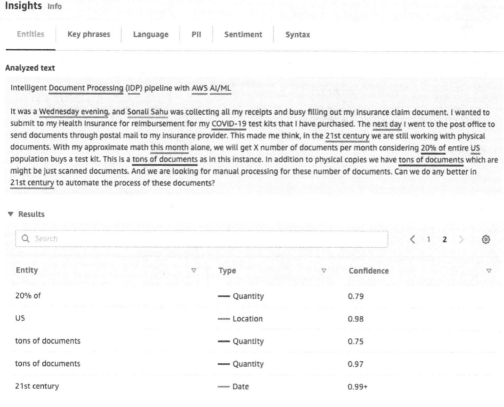

Insights Info

| Entities | Key phrases | Language | PII | Sentiment | Syntax |

Analyzed text

Intelligent Document Processing (IDP) pipeline with AWS AI/ML

It was a Wednesday evening, and Sonali Sahu was collecting all my receipts and busy filling out my insurance claim document. I wanted to submit to my Health Insurance for reimbursement for my COVID-19 test kits that I have purchased. The next day I went to the post office to send documents through postal mail to my insurance provider. This made me think, in the 21st century we are still working with physical documents. With my approximate math this month alone, we will get X number of documents per month considering 20% of entire US population buys a test kit. This is a tons of documents as in this instance. In addition to physical copies we have tons of documents which are might be just scanned documents. And we are looking for manual processing for these number of documents. Can we do any better in 21st century to automate the process of these documents?

▼ Results

Q Search   ‹ 1 **2** › ⚙

Entity	▽	Type	▽	Confidence	▽
20% of		— Quantity		0.79	
US		— Location		0.98	
tons of documents		— Quantity		0.75	
tons of documents		— Quantity		0.97	
21st century		— Date		0.99+	

Figure 4.6 – Amazon Comprehend entities

- **Detecting key phrases**: Amazon Comprehend can detect key phrases in a paragraph of text. To detect key phrases, we will go to the **Key phrases** tab. In English, a key phrase consists of a noun phrase (noun plus modifier) that describes a particular thing. For the text in *Figure 4.3*, you can see the extracted key phrases here:

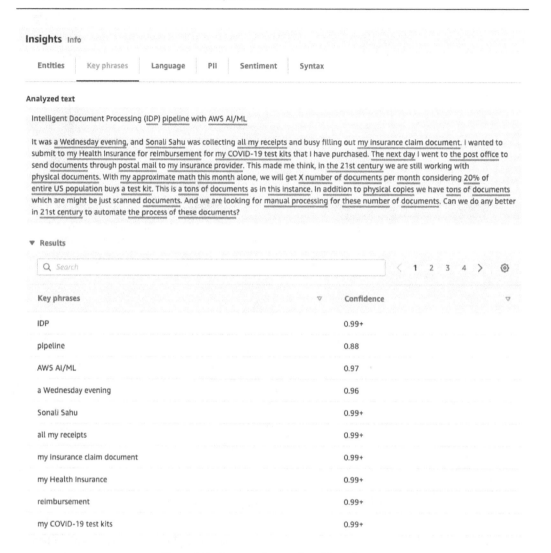

Figure 4.7 – Amazon Comprehend key phrases

At times, we get documents in different languages. We can identify the dominant language of the text.

- **Detecting language:** This is one of the most common requirements for businesses that process documents globally. I was speaking with a college transcripts processing customer, they get college applications and supporting documents from many different countries that can be in different languages, such as Spanish, German, Korean, and more. The first thing to find out is, does your extraction engine support the language in question? For example, Amazon Textract supports Spanish, Italian, German, and more, in addition to English. My recommendation is that you find an **Optical Character Recognition** (**OCR**) engine that supports your language. If your requirement is just to detect the language of some text for further downstream processing or just categorization based on the language, you can use Amazon Comprehend's detect dominant language feature. In *Figure 4.7*, we have gone to the **Language** tab to see the dominant language identified by Amazon Comprehend:

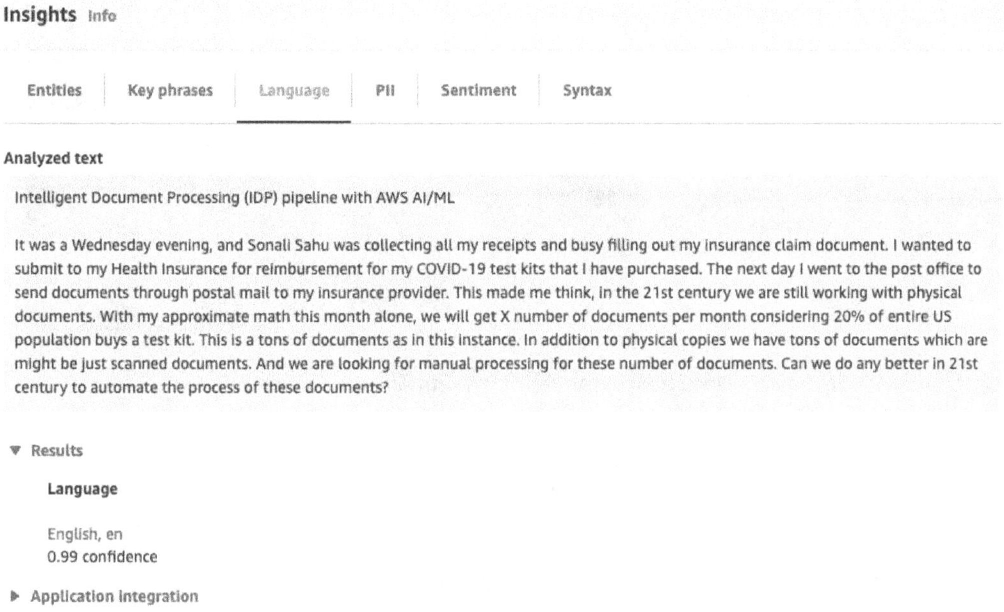

Insights Info

| Entities | Key phrases | Language | PII | Sentiment | Syntax |

Analyzed text

Intelligent Document Processing (IDP) pipeline with AWS AI/ML

It was a Wednesday evening, and Sonali Sahu was collecting all my receipts and busy filling out my insurance claim document. I wanted to submit to my Health Insurance for reimbursement for my COVID-19 test kits that I have purchased. The next day I went to the post office to send documents through postal mail to my insurance provider. This made me think, in the 21st century we are still working with physical documents. With my approximate math this month alone, we will get X number of documents per month considering 20% of entire US population buys a test kit. This is a tons of documents as in this instance. In addition to physical copies we have tons of documents which are might be just scanned documents. And we are looking for manual processing for these number of documents. Can we do any better in 21st century to automate the process of these documents?

▼ Results

Language

English, en
0.99 confidence

▶ Application integration

Figure 4.8 – Amazon Comprehend language detection

Another business-critical feature of Amazon Comprehend is PII detection. We will be covering this in detail in *Chapter 6, Review and Verification of Intelligent Document Processing*.

Now let's see how key features of Amazon Comprehend can be extracted using APIs for the extraction stage of the IDP pipeline.

Understanding document extraction – the IDP extraction stage with Amazon Comprehend

In the preceding example for Amazon Comprehend's extraction, the input required was of the text type.

How can we process documents and extract insights with Amazon Comprehend? For this solution, we will use Amazon Textract in conjunction with Amazon Comprehend for accurate data extraction.

See *Figure 4.9* for an architecture that would serve as the extraction part of the IDP pipeline:

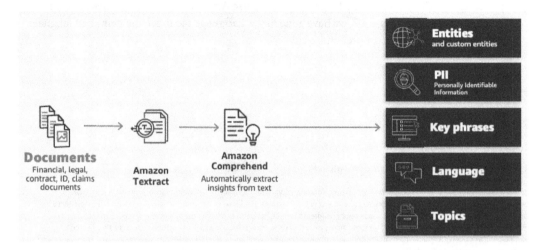

Figure 4.9 – Document extraction stage with Amazon Comprehend

We have walked through the key features of Amazon Comprehend on the AWS Management Console. But we can use Amazon Comprehend APIs to automate extraction programmatically. Now let's walk through some sample code for extracting pre-trained entities from any type of document:

1. Get the `boto3` client for Amazon Textract and Amazon Comprehend:

    ```
    s3=boto3.client('s3')
    textract = boto3.client('textract', region_name=region)
    comprehend=boto3.client('comprehend', region_name=region)
    ```

2. We will use `04detectenttitiesdoc.png`. Let's check out the content of this document:

    ```
    # Document
    documentName = "04detectentitiesdoc.png"
    display(Image(filename=documentName))
    ```

    > Hello Zhang Wei, I am John. Your AnyCompany Financial Services, LLC credit card account 1111-0000-1111-0008 has a minimum payment of $24.53 that is due by July 31st. Based on your autopay settings, we will withdraw your payment on the due date from your bank account number XXXXXX1111 with the routing number XXXXX0000.
    >
    > Your latest statement was mailed to 100 Main Street, Any City, WA 98121.
    > After your payment is received, you will receive a confirmation text message at 206-555-0100. If you have questions about your bill, AnyCompany Customer Service is available by phone at 206-555-0199 or email at support@anycompany.com.

    Figure 4.10 – Sample text for the Amazon Comprehend API

3. Call the Amazon Textract `detect_document_text()` API for the extraction of text from the document. Then, we parse through each line of the JSON response and print the result as follows:

```
#Extract dense Text from scanned document
# Amazon Textract client
textract = boto3.client('textract')
# Read document content
with open(documentName, 'rb') as document:
    imageBytes = bytearray(document.read())
# Call Amazon Textract
response = textract.detect_document_
text(Document={'Bytes': imageBytes})
# Print detected text
text = ""
for item in response["Blocks"]:
    if item["BlockType"] == "LINE":
        print ('\033[94m' +  item["Text"] + '\033[0m')
        text = text + " " + item["Text"]
```

You can see the result of all extracted lines from Amazon Textract in *Figure 4.11*:

```
Hello Zhang Wei, I am John. Your AnyCompany Financial Services, LLC credit card account 1111-
0000-1111-0008 has a minimum payment of $24.53 that is due by July 31st. Based on your
autopay settings, we will withdraw your payment on the due date from your bank account
number XXXXXX1111 with the routing number XXXXX0000.
Your latest statement was mailed to 100 Main Street, Any City, WA 98121.
After your payment is received, you will receive a confirmation text message at 206-555-0100.
If you have questions about your bill, AnyCompany Customer Service is available by phone at
206-555-0199 or email at support@anycompany.com.
```

Figure 4.11 – Lines extracted with Amazon Textract

4. Now let's pass the extracted text in the preceding step to Amazon Comprehend for further insights extraction. We are calling Amazon Comprehend's `detect_entities()` API to extract pre-trained entities:

```
#detect entities
entities = comprehend.detect_entities(LanguageCode="en",
Text=text)
#print(entities)
#Print pre-defined entities of type Person
print("\nPersons identified")
```

```
for item in entities["Entities"]:
    if item["Type"] == "PERSON":
        print(item["Text"])
print("\nOrganization identified")
#print pre-defined entities of type Organization
for item in entities["Entities"]:
    if item["Type"] == "ORGANIZATION":
        print(item["Text"])
print("\nQuantities identified")
#print pre-defined entities of type Quantity
for item in entities["Entities"]:
    if item["Type"] == "QUANTITY":
        print(item["Text"])
print("\nLocation identified")
#print pre-defined entities of type Location
for item in entities["Entities"]:
    if item["Type"] == "LOCATION":
        print(item["Text"])
```

5. We are iterating through the JSON response and filtering for `"PERSON"`, `"ORGANIZATION"`, `"QUANTITY"`, and `"LOCATION"`. The results are as follows:

```
Persons identified
Zhang Wei
John

Organization identified
AnyCompany Financial Services, LLC
AnyCompany Customer Service

Quantities identified
$24.53

Location identified
100 Main Street, Any City, WA 98121
```

Figure 4.12 – Amazon Comprehend results

Here, we leveraged Amazon Textract and Amazon Comprehend for accurate document extraction.

Amazon Comprehend also supports custom entities extraction in addition to pre-trained entities extraction. Now let's look at custom entities extraction and how it can be done accurately with Amazon Comprehend.

Understanding custom entities extraction with Amazon Comprehend

Sometimes, our key business terms don't fall under the category of pre-defined entities. In those cases, we can train our custom entity recognizer to get insights from our document. Amazon Comprehend's custom entity recognition allows you to bring in your own dataset (a list of documents) and train a custom model to extract custom entities from your documents. This is a two-step process:

1. Train an entity recognizer by providing a small, labeled dataset. This entity recognizer uses **automated ML (AutoML)** and transfer learning to train a model based on your training dataset. It also offers evaluation/performance metrics, such as F1 score, precision, and recall. You can start training an Amazon Comprehend custom entity recognizer with single-digit sample documents. I recommend that you check the performance metrics of the trained model and include additional training samples to improve them. Also, you should check your model training for overfitting.

2. Run asynchronous or real-time analysis on the trained model after you have trained a custom entity.

Now let's walk through each step.

We are using **Home Owner Association (HOA)** documents for our sample and extracting custom entities such as "closing date," "seller," "address," and "due amount." HOA documents contain dense text along with additional semi-structured form-based information. Businesses want to extract only information or business terms that are critical to their use case. For example, from an HOA document, a business may want to extract closing dates and addresses for further review and comparison. You can also tag and index documents with such metadata to enable intelligent search. However, adding such search capabilities to the IDP pipeline is out of the scope of this book. Now, we will look into the extraction of custom entities from any type of document.

We will be using the following architecture for our Amazon Comprehend custom entity recognition model:

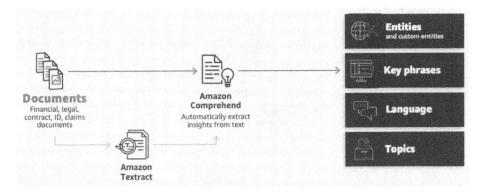

Figure 4.13 – Document extraction with Amazon Comprehend/Amazon Textract

Let's go through the training of our Amazon Comprehend custom entity recognizer.

Training an Amazon Comprehend custom entity recognizer

Amazon Comprehend allows two types of training datasets. The simpler type is the entity list, where your training data consists of text and text type. This is easy to get started with, but for more accurate results, I recommend creating labeled training data with specific offsets. To give additional training context, we can use annotations and training documents with exact bounding boxes by using **Begin Offset** and **End Offset**.

Training dataset

A training dataset teaches your model to recognize entities. **See guidance** [⤴]

Training type

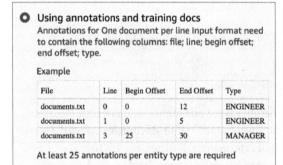
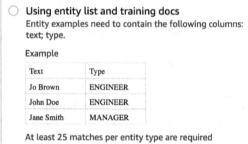

Figure 4.14 – Training dataset selection for Amazon Comprehend

For our example, we are using `entity_list`:

1. Use the CSV file named `entity_list.csv`. We are also printing the custom entities that we want to derive from the documents:

    ```
    entities_df = pd.read_csv('./chapter_4_training/entity_
    list.csv', dtype={'Text': object})
    entities = entities_df["Type"].unique().tolist()
    print(f'Custom entities : {entities}')
    print(f'\nTotal Custom entities: {entities_df["Type"].
    nunique()}')
    display(HTML(entities_df.to_html(index=False)))
    ```

2. Now let's look at the labeled data in *Figure 4.15*:

```
Custom entities : ['CLOSING_DATE', 'SELLER', 'ADDRESS', 'DUE_AMOUNT']
Total Custom entities: 4
```

	Text	Type
	04/11/2022	CLOSING_DATE
	Alonso Weiss	SELLER
7200 Discovery Drive, Chattanooga TN 37416-1757		ADDRESS
	$ 3448.12	DUE_AMOUNT
	04/20/2022	CLOSING_DATE
	Albert Boyd	SELLER
14840 Central Pike Suite 190,Lebanon, TN 37090		ADDRESS

Figure 4.15 – Labeled data for Comprehend custom entities

3. Now let's upload our `entity_list` training dataset to Amazon S3. We are also uploading our training corpus (`entity_training_corpus.csv` `https://github.com/PacktPublishing/Intelligent-Document-Processing-with-AWS-AI-ML-/blob/main/chapter-4/entity_training_corpus.csv`) to the same S3 bucket:

```
#Upload entity list CSV to S3
entities_key='./chapter_4_training/entity_list.csv'
training_data_key='./chapter_4_training/entity_training_
corpus.csv'
data_bucket = "bookidppackt123"
s3=boto3.client('s3')
s3.upload_file(Filename='./chapter_4_training/entity_
list.csv',
                Bucket=data_bucket,
                Key=entities_key)
s3.upload_file(Filename='./chapter_4_training/entity_
training_corpus.csv',
                Bucket=data_bucket,
                Key=training_data_key)
entities_uri = f's3://{data_bucket}/{entities_key}'
training_data_uri = f's3://{data_bucket}/{training_data_
key}'
print(f'Entity List CSV File: {entities_uri}')
print(f'Training Data File: {training_data_uri}')
```

You can see the result for the labeled training dataset in *Figure 4.16*:

```
Entity List CSV File: s3://███████████/./chapter_4_training/entity_list.csv
Training Data File: s3://███████████/./chapter_4_training/entity_training_corpus.csv
```

Figure 4.16 – Labeled data for our custom Comprehend NER model

4. We are calling Comprehend's `create_entity_recognizer()` API to create a custom Comprehend model. We are passing in a training data URI and an entity list URI as inputs:

```
# Create a custom entity recognizer
import sagemaker
entity_recognizer_name = 'Sample-Entity-Recognizer-
IDPBook'
entity_recognizer_version = 'Sample-Entity-Recognizer-
IDPBook-v1'
entity_recognizer_arn = ''
create_response = None
role = sagemaker.get_execution_role()
EntityTypes = [ {'Type': entity} for entity in entities]
try:
    create_response = comprehend.create_entity_
recognizer(
        InputDataConfig={
            'DataFormat': 'COMPREHEND_CSV',
            'EntityTypes': EntityTypes,
            'Documents': {
                'S3Uri': training_data_uri
            },
            'EntityList': {
                'S3Uri': entities_uri
            }
        },
        DataAccessRoleArn=role,
        RecognizerName=entity_recognizer_name,
        VersionName=entity_recognizer_version,
        LanguageCode='en'
    )
    entity_recognizer_arn = create_
response['EntityRecognizerArn']
    print(f"Comprehend Custom entity recognizer created
with ARN: {entity_recognizer_arn}")
```

5. It takes some time to train a custom Comprehend entity recognition model. You can check the status of your trained Comprehend model by going to Amazon Comprehend on the AWS Management Console or running API `DescribeEntityRecognizer`.

Now let's check the performance of our trained model.

Checking the performance of a trained model

Within Amazon Comprehend in the AWS Management Console, click on **Custom entity recognition** and select the custom NER model that we trained in the previous steps. Click on the **Performance** tab to check the performance of the model:

Input & Output	Performance	Endpoints	Tags, VPC & Policy	Application Integration

Version performance Info

Test data source	F1 score	Precision	Recall
Autosplit	94.67	100	89.88

Entity types performance

Custom entity type ▲	F1 score ▽	Precision ▽	Recall ▽	Number of train mentions
ADDRESS	100	100	100	160
CLOSING_DATE	81.63	100	68.96	220
DUE_AMOUNT	100	100	100	160
SELLER	100	100	100	160

Figure 4.17 – Performance metrics for the Comprehend custom entities model

Inference result from the Amazon Comprehend custom entity recognizer

Now our Amazon Comprehend custom entity recognizer is trained. We will follow these steps to inspect the results of our trained custom NER model:

1. Go to Amazon Comprehend in the AWS Management Console. Click on **Analysis jobs** on the left:

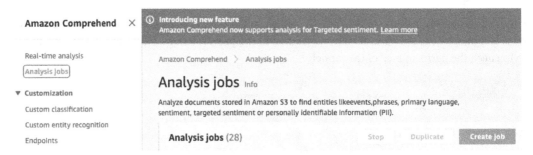

Figure 4.18 – Analysis jobs on the AWS Management Console

2. Then click on **Create job**:

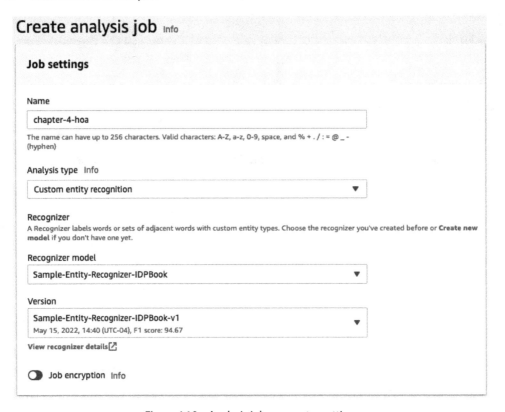

Figure 4.19 – Analysis job parameter setting

3. As the input document, upload `hoa_0.pdf` to your Amazon S3 bucket, and give the path to this file in the Amazon S3 bucket.

Figure 4.20 – Analysis job input data

4. For **Output data location**, give the location of your Amazon S3 bucket.

5. For **Access permissions**, select **Create an IAM Role** and provide the prefix as shown in the following screenshot:

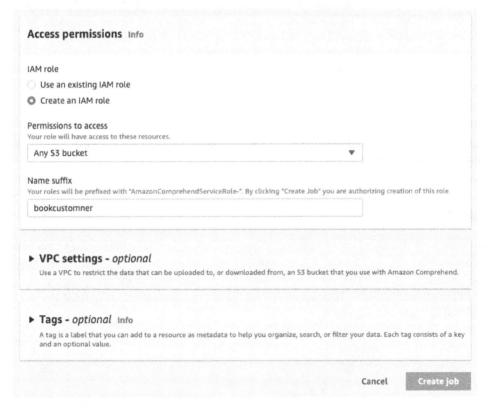

Figure 4.21 – Analysis job access permissions

6. Then click on **Create job**. It will take some time to get the result.

7. Once the analysis job is complete, you can go to your output location bucket and check the result.

    A sample result follows:

```
{"Entities": [{"BeginOffset": 14, "EndOffset": 24,
"Score": 0.9999934435319415, "Text": "04/04/2022",
"Type": "CLOSING_DATE"}], "File": "hoa_0.txt", "Line": 6}
{"Entities": [{"BeginOffset": 8, "EndOffset": 18,
"Score": 0.9999991059312681, "Text": "John Smith",
"Type": "SELLER"}], "File": "hoa_0.txt", "Line": 7}
{"Entities": [{"BeginOffset": 25, "EndOffset": 60,
"Score": 0.9999999850988398, "Text": "28777 Amos Lock,
Markfurt, HI 71418", "Type": "ADDRESS"}], "File": "hoa_0.
txt", "Line": 8}
{"Entities": [{"BeginOffset": 0, "EndOffset": 8, "Score":
0.9999828342970942, "Text": "$ 315.58", "Type": "DUE_
AMOUNT"}], "File": "hoa_0.txt", "Line": 14}
```

8. As you can see, we were able to accurately extract "CLOSING_DATE", "SELLER", "ADDRESS", and "DUE_AMOUNT" from our HOA document. We also got a confidence score. We can leverage confidence scores to set a threshold for the accurate extraction of elements.

We just saw how we can train our own custom Comprehend entity recognition model to extract entities from any type of document.

> **Note**
>
> Comprehend charges for trained models, so I recommend you delete your trained model and any datasets to avoid unexpected costs.

Summary

In this chapter, we discussed core features of Amazon Comprehend, including the extraction of pre-trained entities such as "Person," "Date," and "Location" from text. We then discussed how we can leverage Amazon Comprehend for the document extraction stage of the IDP pipeline. We also discussed how to use Amazon Textract to extract text from a document and pass it to Amazon Comprehend for entity extraction.

We then reviewed the need for custom entities extraction and how to train your own Comprehend custom entity recognizer model. We discussed the two-step process of training a custom entity recognizer and then created an analysis job for custom entities extraction from any type of document.

In the next chapter, we will extend the extraction and enrichment stage of the IDP pipeline using Amazon Comprehend Medical. You will be introduced to the enrichment stage of IDP and discover how to leverage Amazon Comprehend to enrich your documents.

Part 2: Enrichment of Data and Post-Processing of Data

In this part, we will learn about Amazon Comprehend Medical and its application in the enrichment stage of the IDP pipeline. Then, we will focus on how to post-process documents during the IDP pipeline. Moreover, we will learn how to create a secure data pipeline with the right access control. Finally, we will learn how to post-process documents during the IDP pipeline.

This section comprises the following chapters:

- *Chapter 5, Document Enrichment in Intelligent Document Processing*
- *Chapter 6, Review and Verification of Intelligent Document Processing*
- *Chapter 7, Accurate Extraction and Health Insights with Amazon HealthLake*

5

Document Enrichment in Intelligent Document Processing

In the previous chapter, you read how businesses can use Amazon Comprehend to extract insights from any type of document. You learned about document extraction with Amazon Comprehend's entities and custom entities extraction feature. You also learned how to train your own custom to comprehend models with Amazon Comprehend. We also discussed how to use Amazon's **Custom Entity Recognizer** to extract key business terms from documents. We will now dive into the details of the **document enrichment** stage of the **IDP pipeline** and how we can extract and enrich our IDP pipeline with medical insights from medical documents. You will get an understanding of the enrichment stage of IDP by getting health insights with Amazon Comprehend Medical with its features like medical entities extraction, and ontology linking. We will also look into additional document enrichment examples by identifying, and redacting sensitive information in any types of documents. .

We will navigate through the following sections in this chapter:

- Understanding document enrichment
- Learning to use Amazon Comprehend Medical for accurate extraction of medical entities
- Learning to use Amazon Comprehend Medical for medical ontology

Technical requirements

If you have not already done so, please follow the *Technical requirements* section in *Chapter 2, Document Capture and Categorization*.

Understanding document enrichment

To get insights and business value from your documents, you will need to understand dynamic topics and document attributes. You also have a requirement to augment documents, such as redacting sensitive information, translating the extracted data to an additional language as per your needs, or just augmenting documents with inferred metadata. Getting these additional insights is known as the document enrichment stage of the IDP pipeline. During the document enrichment stage, you augment and append your existing document insights with business- or domain-specific context from additional sources. Let's discuss a couple of examples of document enrichment.

For example, your need is to translate documents from English to Spanish to support global customers. You can leverage AWS AI services such as Amazon Translate to translate documents from one language to another. We will use the following architecture for the translation of text in a document.

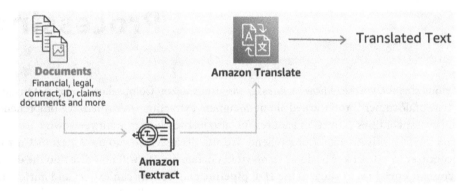

Figure 5.1 – Language translation architecture

In *Figure 5.1*, we are leveraging Amazon Translate to translate text. Amazon Translate is a text translation AI service that offers high-quality and accurate translation. Amazon Translate takes text as input, so we are using Amazon Textract to extract text from a document and pass it in as input to Amazon Translate.

> **Note**
> We recommend including a queue to store the Textract response and make it a decoupled architecture.

Let's check that with a hands-on code sample to translate a text from English into German:

1. Import the required libraries:

    ```
    import boto3
    import uuid
    ```

```
from io import BytesIO
import sys
from IPython.display import Image, display
```

2. Now let's check the sample document:

```
# Document
documentName = "syncdensetext.png"
display(Image(filename=documentName))
```

3. Get the boto3 client for Textract, extract all text as raw bytes from the document, and print all the LINES:

```
# Read document content
with open(documentName, 'rb') as document:
    imageBytes = bytearray(document.read())
# Call Amazon Textract
response = textract.detect_document_
text(Document={'Bytes': imageBytes})
# Print detected text
for item in response["Blocks"]:
    if item["BlockType"] == "LINE":
        print (item["Text"])
```

4. Now pass all Textract LINES to Amazon Translate to translate the English text to German:

```
# Amazon Translate client
translate = boto3.client('translate')
print ('')
for item in response["Blocks"]:
    if item["BlockType"] == "LINE":
        print ('\033[94m' +  item["Text"] + '\033[0m')
        result = translate.translate_
text(Text=item["Text"], SourceLanguageCode="en",
TargetLanguageCode="de")
        print ('\033[92m' + result.get('TranslatedText')
+ '\033[0m')
    print ('')
```

5. You can see the sample translated text here:

```
It was a Wednesday evening, and I was collecting all my receipts and busy filling out my
Es war ein Mittwochabend, und ich sammelte alle meine Quittungen ab und füllte meine

insurance claim document. I wanted to submit to my Health Insurance for reimbursement for
Dokument für Versicherungsansprüche. Ich wollte bei meiner Krankenversicherung eine Erstattung beantragen

my COVID-19 test kits that I have purchased. The next day I went to the post office to send
meine COVID-19-Testkits, die ich gekauft habe. Am nächsten Tag ging ich zur Post, um zu schicken

documents through postal mail to my insurance provider. This made me think, in the 21 st
Dokumente per Post an meine Versicherung. Das brachte mich zum Nachdenken, im 21.

century we are still working with physical documents. With my approximate math this month
Jahrhundert arbeiten wir immer noch mit physischen Dokumenten. Mit meiner ungefähren Mathematik in diesem Monat
```

Figure 5.2 – Sample translated text output

Another common example of the document enrichment stage is to identify sensitive information such as PII/PHI in your document, and redact that sensitive information to meet your regulatory and compliance requirements. For example, during mortgage processing application we have documents with information such as Name, SSN information. We want to not only identify these sensitive information during Document Enrichment stage, but also want to redact them and set right access control. We will cover this feature in more detail in *Chapter 6, Review and Verification of Intelligent Document Processing*, under the *Post processing to handle sensitive data* section.

Now let's look into a healthcare industry document processing use case of the document enrichment stage. While processing healthcare claims, at times, we need to refer to doctors' notes to verify the medical condition mentioned in the claims form. Additional documents such as doctors' notes are requested for further processing. We get a raw doctor's note, and we will extract and infer medical information such as medication and medical condition, leveraging AWS AI services. This process often includes skilled medical professionals. But by automating this process, we can leverage skilled medical people to deliver better patient care. To achieve this, we need medical context, metadata, attributes, and domain-specific knowledge. This is an example of the enrichment stage of the IDP pipeline.

In the following figure, you can see the document enrichment stage in the IDP pipeline leveraging Amazon Comprehend and Amazon Comprehend Medical.

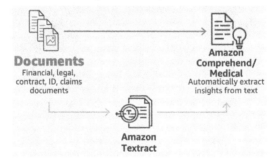

Figure 5.3 – Intelligent document processing – extraction/enrichment

So now, let's dive deep into Amazon Comprehend Medical.

Learning to use Amazon Comprehend Medical for accurate extraction of medical entities

Amazon Comprehend Medical is a **Natural Language Processing** (**NLP**) service for extracting medical insights from health data. Like other AI services such as Amazon Textract, Amazon Comprehend Medical has been pretrained by a machine learning model to understand the medical context. It is a managed solution, meaning you call Amazon Comprehend Medical's API and pass in your input, and you will get your response back. To leverage Comprehend, no ML experience is required. You can call its **Application Programming Interfaces** (**APIs**) to leverage the pre-trained ML model behind the scenes. Amazon Comprehend provides APIs that you can call through the serverless environment without the need to manage any kind of infrastructure. Amazon Comprehend Medical is a HIPAA-eligible service, meeting any healthcare customer's compliance standard.

Now let's check the core features and functionalities of Amazon Comprehend Medical.

Amazon Comprehend Medical

Amazon Comprehend Medical is an NLP machine learning service trained in the medical context to get health insights from data. Amazon Comprehend Medical enables you to examine unstructured data, for example, any type of unstructured text. It can help you gain medical insights from its content by using a number of pretrained models.

In *Figure 5.4*, you can see some of the key features of Amazon Comprehend Medical for the document extraction/enrichment phase of the intelligent document processing pipeline.

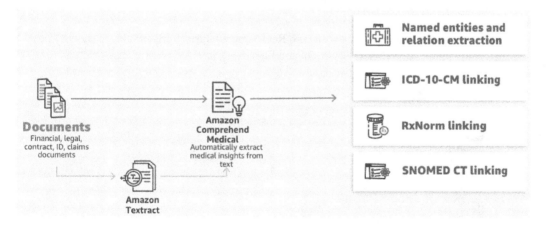

Figure 5.4 – Amazon Comprehend Medical key features

Now Let's have a walk-through of these features on the AWS console for Amazon Comprehend Medical:

1. Go to Amazon Comprehend Medical. Click on **Launch real-time analysis**:

Figure 5.5 – Amazon Comprehend Medical console

2. We will use the following sample text to analyze all of the features of Amazon Comprehend Medical available through the AWS console:

```
Patient is a 85 yo woman, school bus driver with past
medical history that includes
    - status post cardiac catheterization in April 2021.
She presents today with palpitations and chest pressure.
Meds : Vyvanse 50 mgs po at breakfast daily,
            Clonidine 0.2 mgs -- 1 and 1 / 2 tabs po qhs
Lungs : clear
Heart : Regular rhythm
Follow-up as scheduled
```

3. Copy the preceding text and insert it into **Real-time analysis → Input text**, as shown in the preceding figure, then click on **Built-in**, and then **Analyze**:

Figure 5.6 – Amazon Comprehend Medical real-time analysis

4. Scroll down to see the insights. The following figure shows the UI representation of medical insights from the text:

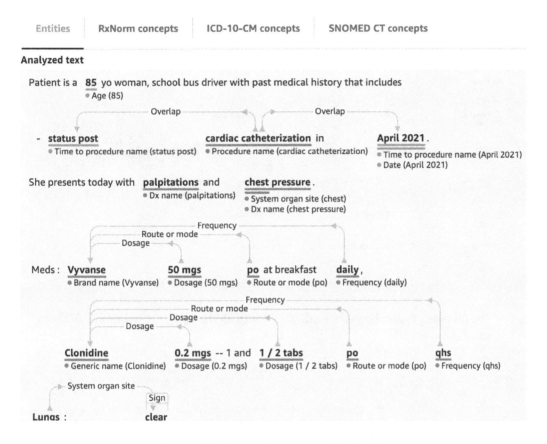

Figure 5.7 – Amazon Comprehend Medical UI

- The following figure shows the medical conditions inferred from the preceding text:

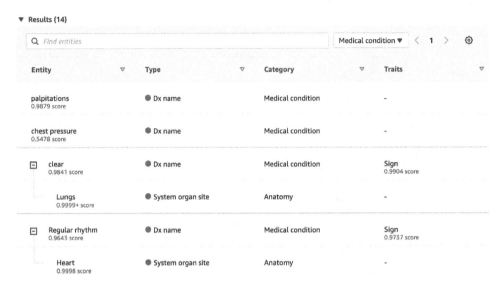

Figure 5.8 – Amazon Comprehend Medical conditions

- The following figure shows the medication lists with the dosage, router mode, and brand/generic name:

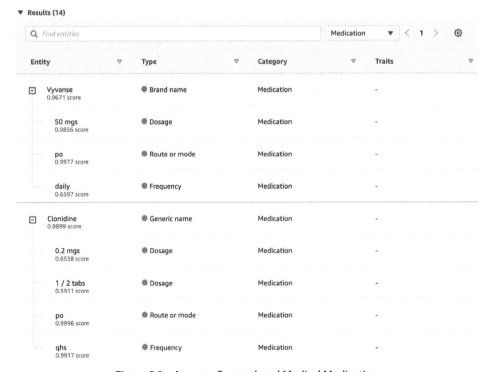

Figure 5.9 – Amazon Comprehend Medical Medication

- The following figure shows the test treatment procedures inferred from the text:

Figure 5.10 – Amazon Comprehend Medical Test treatment procedures

- The following figure shows the anatomy inferred from the text:

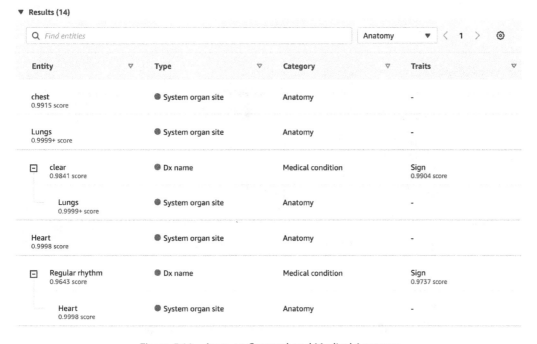

Figure 5.11 – Amazon Comprehend Medical Anatomy

- The following figure shows the time expression of the medical events inferred from the text.

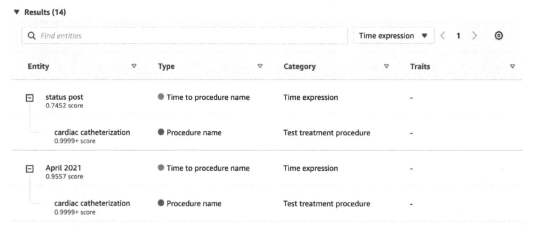

Figure 5.12 – Amazon Comprehend Medical Time expression

- The following figure shows the RxNorm concepts inferred from the text:

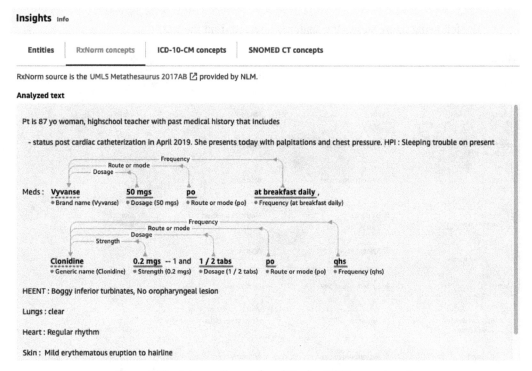

Figure 5.13 – Amazon Comprehend Medical RXNorm concepts

- The following figure shows the ICD-10-CM concepts inferred from the text:

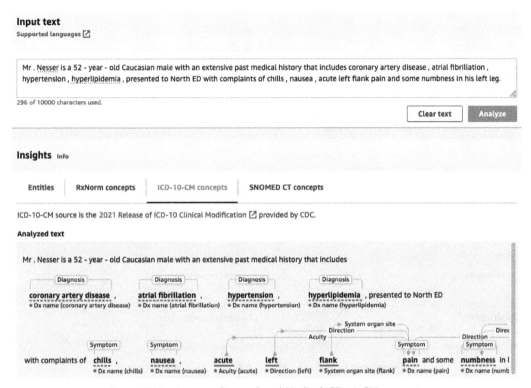

Figure 5.14 – Amazon Comprehend Medical ICD-10-CM concepts

Here, you can also check the sample inferred ICD-10-CM codes from the text:

coronary artery disease

Top inferred concepts

I25.10   Atherosclerotic heart disease of native coronary
            artery without angina pectoris
            Score: 0.9260

I25.119   Atherosclerotic heart disease of native coronary
            artery with unspecified angina pectoris
            Score: 0.8527

I25.110   Atherosclerotic heart disease of native coronary
            artery with unstable angina pectoris
            Score: 0.8507

Z86.79   Personal history of other diseases of the
            circulatory system
            Score: 0.8274

I25.41   Coronary artery aneurysm
            Score: 0.8204

▶ **More information**

atrial fibrillation

Top inferred concepts

I48.91   Unspecified atrial fibrillation
            Score: 0.9246

I48.0   Paroxysmal atrial fibrillation
            Score: 0.8910

I48.2   Chronic atrial fibrillation
            Score: 0.8813

I48.1   Persistent atrial fibrillation
            Score: 0.8529

I48.92   Unspecified atrial flutter
            Score: 0.8246

▶ **More information**

hypertension

Top inferred concepts

I10   Essential (primary) hypertension
            Score: 0.8898

I11.0   Hypertensive heart disease with heart failure
            Score: 0.8686

I11.9   Hypertensive heart disease without heart failure
            Score: 0.8337

I13.0   Hypertensive heart and chronic kidney disease
            with heart failure and stage 1 through stage 4
            chronic kidney disease, or unspecified chronic
            kidney disease
            Score: 0.8313

I12.9   Hypertensive chronic kidney disease with stage 1
            through stage 4 chronic kidney disease, or
            unspecified chronic kidney disease
            Score: 0.8286

hyperlipidemia

Top inferred concepts

E78.5   Hyperlipidemia, unspecified
            Score: 0.9294

E78.2   Mixed hyperlipidemia
            Score: 0.8411

Z83.438   Family history of other disorder of lipoprotein
            metabolism and other lipidemia
            Score: 0.8287

E78.1   Pure hyperglyceridemia
            Score: 0.8258

E78.00   Pure hypercholesterolemia, unspecified
            Score: 0.8179

Figure 5.15 – Amazon Comprehend Medical ICD-10-CM codes

- The following figure shows the SNOMED CT concepts inferred from the text:

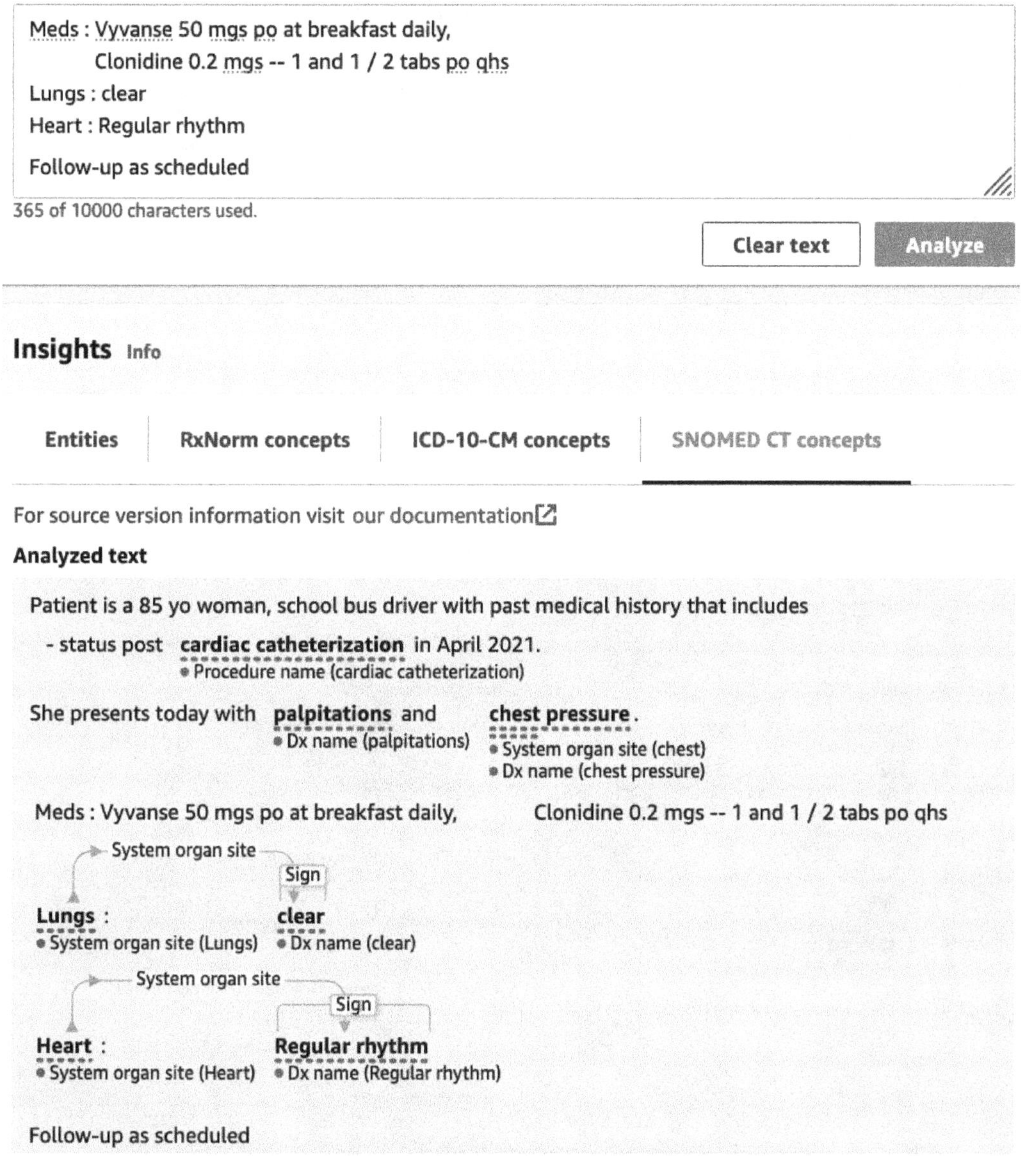

Meds : Vyvanse 50 mgs po at breakfast daily,
Clonidine 0.2 mgs -- 1 and 1 / 2 tabs po qhs
Lungs : clear
Heart : Regular rhythm

Follow-up as scheduled

365 of 10000 characters used.

Clear text   Analyze

Insights Info

| Entities | RxNorm concepts | ICD-10-CM concepts | SNOMED CT concepts |

For source version information visit our documentation 🗗

Analyzed text

Patient is a 85 yo woman, school bus driver with past medical history that includes

- status post **cardiac catheterization** in April 2021.
  • Procedure name (cardiac catheterization)

She presents today with **palpitations** and   **chest pressure** .
                              • Dx name (palpitations)   • System organ site (chest)
                                                            • Dx name (chest pressure)

Meds : Vyvanse 50 mgs po at breakfast daily,   Clonidine 0.2 mgs -- 1 and 1 / 2 tabs po qhs

➤ System organ site ┐
                    ┌──────┐
                    │ Sign │
                    └──────┘
                        ▼
Lungs :   **clear**
• System organ site (Lungs) • Dx name (clear)

➤ System organ site ┐
                    ┌──────┐
                    │ Sign │
                    └──────┘
                        ▼
Heart :   **Regular rhythm**
• System organ site (Heart) • Dx name (Regular rhythm)

Follow-up as scheduled

Figure 5.16 – Amazon Comprehend Medical SNOMED CT concepts

Here, you can check the inferred possible SNOMED CT codes from the text:

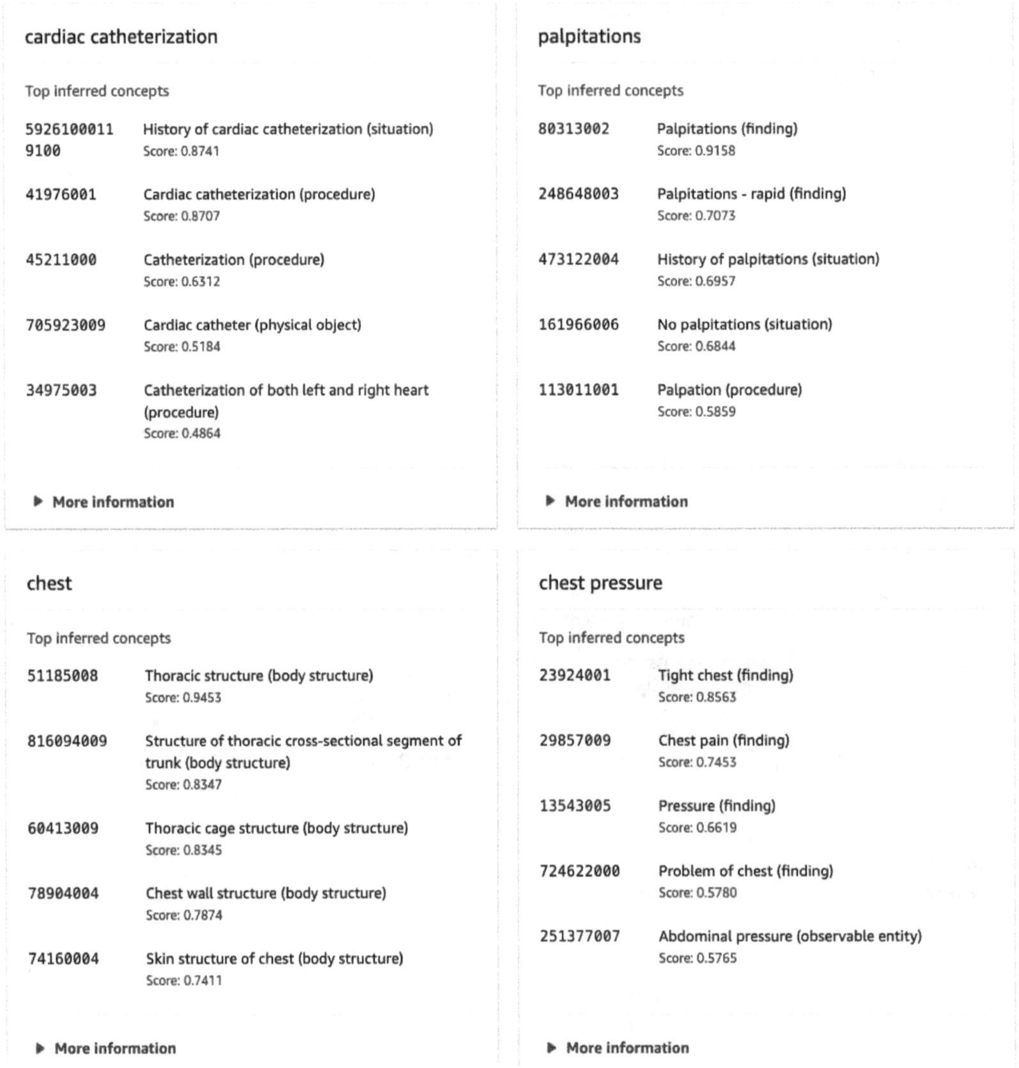

cardiac catheterization

Top inferred concepts

59261000119100	History of cardiac catheterization (situation) Score: 0.8741
41976001	Cardiac catheterization (procedure) Score: 0.8707
45211000	Catheterization (procedure) Score: 0.6312
705923009	Cardiac catheter (physical object) Score: 0.5184
34975003	Catheterization of both left and right heart (procedure) Score: 0.4864

▶ **More information**

palpitations

Top inferred concepts

80313002	Palpitations (finding) Score: 0.9158
248648003	Palpitations - rapid (finding) Score: 0.7073
473122004	History of palpitations (situation) Score: 0.6957
161966006	No palpitations (situation) Score: 0.6844
113011001	Palpation (procedure) Score: 0.5859

▶ **More information**

chest

Top inferred concepts

51185008	Thoracic structure (body structure) Score: 0.9453
816094009	Structure of thoracic cross-sectional segment of trunk (body structure) Score: 0.8347
60413009	Thoracic cage structure (body structure) Score: 0.8345
78904004	Chest wall structure (body structure) Score: 0.7874
74160004	Skin structure of chest (body structure) Score: 0.7411

▶ **More information**

chest pressure

Top inferred concepts

23924001	Tight chest (finding) Score: 0.8563
29857009	Chest pain (finding) Score: 0.7453
13543005	Pressure (finding) Score: 0.6619
724622000	Problem of chest (finding) Score: 0.5780
251377007	Abdominal pressure (observable entity) Score: 0.5765

▶ **More information**

Figure 5.17 – Amazon Comprehend Medical SNOMED CT codes

Now let's walk through the sample code to extract pre-trained medical entities from any type of document:

1. Import the required libraries:

```
import boto3
import json
```

```
import re
import csv
import sagemaker
from sagemaker import get_execution_role
from sagemaker.s3 import S3Uploader, S3Downloader
import uuid
import time
import io
from io import BytesIO
import sys
from pprint import pprint
from IPython.display import Image, display
from PIL import Image as PImage, ImageDraw
```

2. Now let's check the sample provider note:

```
# Document
documentName = "doctornotes1.png"
display(Image(filename=documentName))
```

Here, you can check the raw doctor's note sample document:

The patient is an 86-year-old female admitted for evaluation of abdominal pain and bloody stools. The patient has colitis and also diverticulitis, undergoing treatment. During the hospitalization, the patient complains of shortness of breath, which is worsening. The patient underwent an echocardiogram, which shows severe mitral regurgitation and also large pleural effusion. This consultation is for further evaluation in this regard. As per the patient, she is an 86-year-old female, has limited activity level. She has been having shortness of breath for many years. She also was told that she has a heart murmur, which was not followed through on a regular basis.

Figure 5.18 – Doctor's note sample

3. Get the boto3 client for Textract and extract all text from the document as raw bytes:

```
# process using image bytes
def calltextract(documentName):
    client = boto3.client(service_name='textract',
        region_name= 'us-east-1',
        endpoint_url='https://textract.us-east-1.
amazonaws.com')
```

```
        with open(documentName, 'rb') as file:
                img_test = file.read()
                bytes_test = bytearray(img_test)
                print('Image loaded', documentName)
        # process using image bytes
        response = client.analyze_document(Document={'Bytes':
    bytes_test}, FeatureTypes=['FORMS'])
        return response
```

4. Call the Textract method defined in the preceding code block:

```
response= calltextract(documentName)
print(response)
```

5. Let's print the Amazon Textract response. Let's parse through the JSON to print each LINE:

```
# Print detected text
text = ""
for item in response["Blocks"]:
    if item["BlockType"] == "LINE":
        text = text + " " + item["Text"]
print(text)
```

6. Now let's pass all the LINES extracted previously to Comprehend Medical with the following code. Get the boto3 client for Comprehend Medical and call Comprehend Medical's detect_ entities_v2 () method.

7. We are also parsing Comprehend Medical's JSON response to get "MEDICAL_CONDITION", "ANATOMY", and "TEST_TREATMENT_PROCEDURE":

```
# Call Comprehend Medical
comprehendmedical = boto3.client(service_
name='comprehendmedical')
# Detect medical entities
cm_json_data =  comprehendmedical.detect_entities_
v2(Text=text)
print("\nMedical Condition\n========")
for entity in cm_json_data["Entities"]:
    if entity["Category"] == "MEDICAL_CONDITION":
```

```
        print(entity["Text"])
print("\nAnatomy\n========")
for entity in cm_json_data["Entities"]:
    if entity["Category"] == "ANATOMY":
        print(entity["Text"])
print("\nTest Treatment Procedure\n========")
for entity in cm_json_data["Entities"]:
    if entity["Category"] == "TEST_TREATMENT_PROCEDURE":
        print(entity["Text"])
```

We just looked at sample code to extract medical insights from medical notes. Now let's check how we can use Amazon Comprehend Medical to infer medical ontologies.

Learning to use Amazon Comprehend Medical for medical ontology

What is medical ontology linking? It is the method of identifying medical information and mapping it to standard medical codes and concepts. For example, medical conditions are linked to ICD-10-CM codes. Moreover, medications are mapped to RxNorm codes. Also, Amazon Comprehend Medical results infer SNOMED CT codes to provide medical insights, conditions, affected anatomy, test treatments, and procedures. Amazon Comprehend Medical also supports entity traits. For example, "Patient refused to take medication" has a negation entity trait.

Now let's see an example with a sample prescription document:

1. Use the following code to display the sample prescription document:

    ```
    documentName = "prescription.png"
    display(Image(filename=documentName))
    ```

    You can see the document in the following figure:

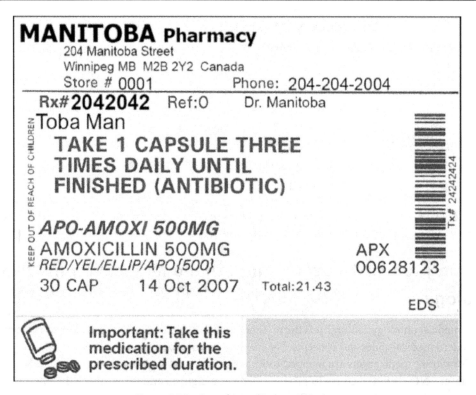

Figure 5.19 – Sample medical prescription

2. Get raw text from the prescription document by calling it into Amazon Textract. We have already defined the `calltextract()` method. So, get the response and parse it to get all the LINES:

```
response= calltextract(documentName)
#print(response)
# Print detected text
text = ""
for item in response["Blocks"]:
    if item["BlockType"] == "LINE":
        text = text + " " + item["Text"]
print(text)
```

Here, you can check the LINES of the prescription document:

```
Image loaded prescription.png
 MANITOBA Pharmacy 204 Manitoba Street Winnipeg MB M2B 2Y2 Canada Store # 0001 Phone: 204-204-2004 Rx#2042042 Ref:0
Dr. Manitoba Toba Man TAKE 1 CAPSULE THREE TIMES DAILY UNTIL FINISHED (ANTIBIOTIC) APO-AMOXI 500MG AMOXICILLIN 500M
G APX RED/YEL/ELLIP/APO{500} 00628123 30 CAP 14 Oct 2007 Total:21.43 EDS D Important: Take this medication for the
prescribed duration.
```

Figure 5.20 – LINES of prescription document

3. Get the Comprehend Medical botot3 client. Pass the Textract raw output to the Amazon Comprehend Medical detect_entities_v2() method and extract "MEDICATION": from the JSON response:

```
# Call Comprehend Medical
comprehendmedical = boto3.client(service_
name='comprehendmedical')
# Detect medical entities
cm_json_data = comprehendmedical.detect_entities_
v2(Text=text)
print("MEDICATION LIST\n")
for entity in cm_json_data["Entities"]:
    if entity["Category"] == "MEDICATION":
        print(entity["Text"])
        for key in entity:
            Attribute_List = []
            if key == 'Attributes':
                for r in entity[key]:
                    Attribute_List.
append(r['Type']+':'+r['Text'])
        print(str(Attribute_List))
```

Let's print the medication result as follows:

```
MEDICATION LIST

APO-AMOXI
['DOSAGE:500MG', 'DOSAGE:500MG', 'FORM:CAP']
AMOXICILLIN
['DOSAGE:500MG', 'FORM:CAP']
```

Figure 5.21 – Medication and dosage from prescription document

4. Call Comprehend Medical's infer_rx_norm() and parse through the JSON response to get the RXNorm code:

```
cm_json_data = comprehendmedical.infer_rx_
norm(Text=text)
print("\n RxNorm Code\n========")
for entity in cm_json_data["Entities"]:
    for rx in entity["RxNormConcepts"]:
        print(rx['Code'])
```

Print the RXNorm result as follows:

```
RxNorm Code
========
1314210
900048
852494
900052
1314330
308191
308212
308192
723
1721474
```

Figure 5.22 – Inferred RXNorm with Amazon Comprehend Medical

5. Now let's look into the code sample to extract ICD-10-CM codes from the text. We are using the doctornotes1.png document, as mentioned in the previous example (*Figure 4.19*). We extract raw text from the document and extract all the LINES from it:

```
# Document
documentName = "doctornotes1.png"
display(Image(filename=documentName))
response= calltextract(documentName)
#print(response)
# Print detected text
text = ""
for item in response["Blocks"]:
    if item["BlockType"] == "LINE":
        text = text + " " + item["Text"]
print(text)
```

6. We pass the extracted text from the previous text to Comprehend Medical's infer_icd10_ cm() method as follows. This API returns a list of ICD-10-CM codes. Amazon Comprehend Medical also returns a confidence score for each element that it detects. We are using a threshold of 93% and filtering all relevant icd10 codes equal to or more than the defined threshold:

```
cm_json_data =  comprehendmedical.infer_icd10_
cm(Text=text)
print("\n Medical coding\n========")
for entity in cm_json_data["Entities"]:
    for icd in entity["ICD10CMConcepts"]:
```

```
if (icd["Score"] >= 0.93):
    code = icd["Code"]
    print(code)
```

Let's check the result:

```
Medical coding
========
K92.1
```

Figure 5.23 – Medical coding from Amazon Comprehend Medical

Let's summarize this chapter next.

Summary

In this chapter, we discussed the document enrichment stage of intelligent document processing with medical insights to augment our document processing with additional insights. We introduced Amazon Comprehend Medical and dove deep into its core features, extracting medical insights such as medical conditions, medication, affected anatomy, time expressions, entity traits, and more from text. We also discussed leveraging Amazon Textract to extract text from medical documents and then passing it to Amazon Comprehend Medical for medical entity extraction. This helps to build a document extraction/enrichment stage for IDP.

We then gave a high-level overview of medical ontology linking and reviewed the need for it. We then looked into the implementation of Amazon Comprehend Medical's ontology linking to extract ICD-10-CM, RXNorm, and SNOMED CT codes from medical documents.

In the next chapter, we will extend the extraction and enrichment stage of the document processing pipeline with Amazon HealthLake. Moreover, we will dive into how we can integrate a HealthLake Data Store with our intelligent document processing pipeline.

Review and Verification of Intelligent Document Processing

In the previous chapter, you understood the requirements for the enrichment process in **Intelligent Document Processing (IDP)**. Often in healthcare industries, there is a requirement to derive medical insights to augment the document processing pipeline. We looked into Amazon Comprehend Medical and its features to derive medical insights for accurate document processing. We will now dive into the detailed post-processing stage of IDP. We will see how the **Review and Verification** steps can be automated by AWS AI services. We will also discuss the requirement and need to have human review and verification options for sensitive, business-critical, or accurate information processing in IDP. We will navigate through the following sections in this chapter:

- Learning post-processing for a completeness check

- Post-processing sensitive data

- Learning about the document review process with human-in-the-loop

Technical requirements

For this chapter, you will need access to an AWS account. Before getting started, we recommend that you create an AWS account by referring to AWS account setup and Jupyter notebook creation steps as mentioned in Technical requirement section in *chapter 2, Document Capture and Categorization*. You can find Chapter-6 code sample in GitHub: `https://github.com/PacktPublishing/ Intelligent-Document-Processing-with-AWS-AI-ML-/tree/main/chapter-6` . Also recommend to check availability of AI service in your AWS regions before using it.

Learning post-processing for a completeness check

Before diving deeper into the implementation of how to use post-processing for a completeness check, first, let's understand the requirements of the Review and Verification stage of the IDP pipeline. In the previous stages of IDP, we discussed how to extract data from documents. Looking inside the documents, we can validate that the key fields needed to process documents meet the accuracy standards set by the business requirements. Most of the time, business uses simple business rules such as whether key fields such as **Name** or **ID** are not empty. For example, in a claims form, you want to make sure the **Insured ID** is always filled in for it to be processed promptly. While we used relatively simple rules in the previous example, you have the ability to construct more complex rules based on your business needs – as an example, reimbursement of over $100,000 (or for X amount) always requires human attention or an additional review. You may also want to cross-check fields between documents. For example, if the total amount is more than X amount, you may ask users to submit additional supporting documents such as invoices or individual receipts for cross-verification. The more we can automate the process, the more time and human effort is required to process documents we can save.

Here, in *Figure 6.1*, is a sample architecture for processing a document using Amazon Textract, and then using business rules to automate the completeness check, as well as the cross-verification process for accurate document processing. We are using the serverless compute AWS Lambda to use our business logic to automate the validation process. For large-scale processing, follow a decoupled architecture with SQS, as mentioned in *Chapter 3, Accurate Document Extraction with Amazon Textract*:

Financial and legal
documents, ID, and
claims documents

**Amazon
Textract**

AWS Lambda
Business Logic for
completeness checks

Figure 6.1 – The completeness check in an IDP pipeline

Let's dive into the implementation of the completeness check in an IDP pipeline.

For this implementation, we are using a sample expense report. Our goal is to check whether the cost for each line item field is accurately filled in. Moreover, any expense total value of more than $2,000 needs to go through an additional review process:

> **Note:**
> For full code walk-through check https://github.com/PacktPublishing/
> Intelligent-Document-Processing-with-AWS-AI-ML-/blob/main/
> chapter-6/postprocessing-06(1).ipynb

1. Import the required libraries. Again, we will be using the `boto3` library for the implementation:

```
import boto3
import sagemaker
```

```
import os
from io import BytesIO
from PIL import Image
from IPython.display import Image, display, JSON, IFrame
from trp import Document
from PIL import Image as PImage, ImageDraw
import time
from IPython.display import IFrame
```

2. Check the "expense.png" image with the following code sample:

```
# Document
documentName = "expense.png"
display(Image(filename=documentName))
```

You can see the screenshotted image in the following figure:

EXPENSE REPORT A
REPORT NUMBER: 35678-9

Expense Report				
Expense Description	Type	Date	Merchant Name	Amount (USD)
Furniture (Desks and Chairs)	Office Supplies	5/10/1019	Merchant One	1500.00
Team Lunch	Food	5/11/2019	Merchant Two	100.00
Team Dinner	Food	5/12/2019	Merchant Three	300.00
Laptop	Office Supplies	5/13/2019	Merchant Three	200.00
			Total	2100.00

Figure 6.2 – The expense report

3. Now, let's extract all elements from the expense report. We will use Amazon Textract for accurate extraction.

4. We get a boto3 client for Amazon Textract. Take all the image bytes out of the expense report. We are calling Amazon Textract's synchronous analyze_document() API. This is an example document with a TABLE type, so we are calling into Textract's API with Feature_Type set to TABLE to extract all the table cells, columns, and rows from the document:

```
# Call Amazon Textract
# Read document content
with open(documentName, 'rb') as document:
    imageBytes = bytearray(document.read())
# Amazon Textract client
```

```
textract = boto3.client('textract')
response = textract.analyze_document(
    Document={'Bytes': imageBytes},
    FeatureTypes=["TABLES"])
```

5. After extracting all the table elements, we are using our custom business rules for the completeness check and additional verification. We are extracting all `expense_values`. We can extend the logic to make sure `expense_value` isn't empty. We are checking for `expense_value` that are greater than $1,000. If a `expense_value` is greater than $1,000, we throw a warning to prompt the user for additional review.

    You can check the following code sample to automate the aforementioned process:

```
doc = Document(response)
warning = ""
for page in doc.pages:
    for table in page.tables:
        for r, row in enumerate(table.rows):
            itemName  = ""
            for col, cell in enumerate(row.cells):
                if(col == 0):
                    itemName = cell.text
                elif(col == 4 and isFloat(cell.text)):
                    expense_value = float(cell.text)
                    if(expense_value > 1000):
                        warning += "{} amount is greater
than $1000.".format(itemName)
if(warning):
    print("\nReview required:\n===================\n" +
warning)
```

6. Here (in *Figure 6.3*) you can see the screenshot of the output. In our expense report, the amount is greater than $1,000, so the **Review required** warning is prompted:

```
Review required:
====================
Furniture (Desks and Chairs)  amount is greater than $1000. amount is greater than $1000.
```

Figure 6.3 – The Review and Validation output

In the preceding code sample, we implemented the completeness check and the validation of the expense report. Now, let's check how to handle sensitive information in the post-processing stage of the IDP pipeline.

Post-processing sensitive data

When processing or sharing documents containing **personally identifiable information** (**PII**) or **protected health information** (**PHI**), it is of the utmost importance that that information is safely removed or redacted to maintain the confidentiality of the beneficiary. Then, what is considered PII information? PII data (as per https://www.dol.gov/general/ppii) is *"any representation of information that permits the identity of an individual to whom the information applies to be reasonably inferred by either direct or indirect means."* Further, PII is defined as information of the following kinds:

- That directly identifies an individual (e.g. name, address, social security number, or another identifying number, code, telephone number, or email address)

- Using which an agency intends to identify specific individuals in conjunction with other data elements – as in, indirect identification.

Also, data privacy and security definitions vary, so it is recommended to check your organization's privacy policy. Companies across industries are looking for ways to redact sensitive information from documents. AWS has AI services to help you automate the identification and redaction of sensitive information from any type of document.

Before diving into a sample implementation, let's check out the Amazon Comprehend PII detection and Amazon Comprehend Medical PHI detection features.

PII detection and redaction in Amazon Comprehend

Amazon Comprehend can identify or redact PII information in English text. With PII detection, you can locate and redact PII entities in the text. To locate PII entities, you can leverage Comprehend's synchronous or asynchronous APIs, but to redact PII entities, you can only use Comprehend's asynchronous batch job API.

Now, let's check out Amazon Comprehend PII detection on AWS Console:

1. Go to **Amazon Comprehend** on the **AWS Console**, as shown here, and click **Launch Amazon Comprehend**:

Figure 6.4 – Amazon Comprehend on the AWS Console

2. Click **Real-time analysis** and click on the **PII** tab. In the following sample, we have given a sample text (`https://github.com/PacktPublishing/Intelligent-Document-Processing-with-AWS-AI-ML-/blob/main/chapter-6/sample-comprehend.txt`) and the screenshot shows the PII information in the text:

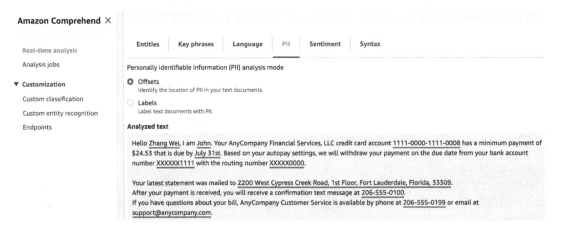

Figure 6.5 – Offsets in Amazon Comprehend PII

3. The PII entities extracted from the preceding text are shown in the following list:

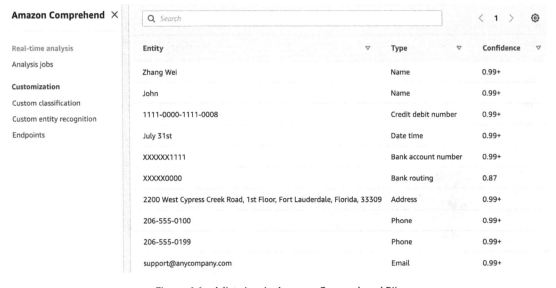

Figure 6.6 – A list view in Amazon Comprehend PII

4. Comprehend can present PII entities as offsets or you may decide to just get the labels of the PII entities. In the following sample figure, a sample output of PII element labels is given by Comprehend:

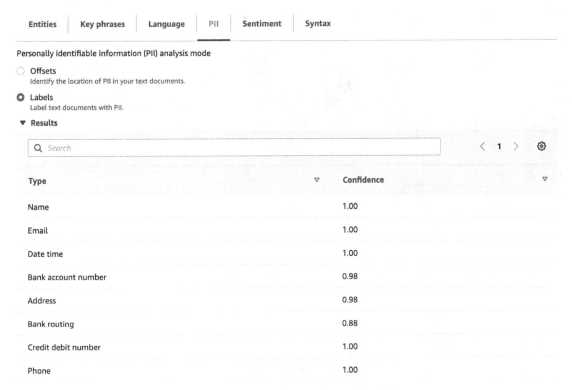

Figure 6.7 – Labels in Amazon Comprehend PII

We have just learned about Amazon Comprehend PII detection on AWS Console. Now, let's check out Comprehend Medical PHI detection.

Amazon Comprehend Medical for PHI information

Amazon Comprehend Medical can detect PHI information from text. It supports a separate API, DetectPHI, for detecting PHI elements in a text. What does it consider a PHI element? For more information, see *Health Information Privacy* (https://www.hhs.gov/hipaa/for-professionals/privacy/special-topics/de-identification/index.html) on the *U.S. Government Health and Human Services* website. Under the HIPAA act, PHI that is based on a list of 18 identifiers must be treated with special care.

Let's check PHI in Amazon Comprehend Medical on the AWS Console:

1. Go to **Amazon Comprehend Medical** on the **AWS Console** and click **Launch real-time analysis**:

Figure 6.8 – Amazon Comprehend Medical on the AWS Console

2. In **Real-time analysis**, we have the following text sample (`https://github.com/ PacktPublishing/Intelligent-Document-Processing-with-AWS-AI-ML-/ blob/main/chapter-6/sample-cm.txt`):

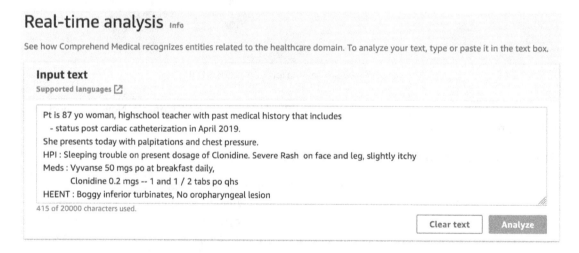

Figure 6.9 – Sample text in Amazon Comprehend Medical

3. You can see the PHI information extracted from the preceding text in the following figure:

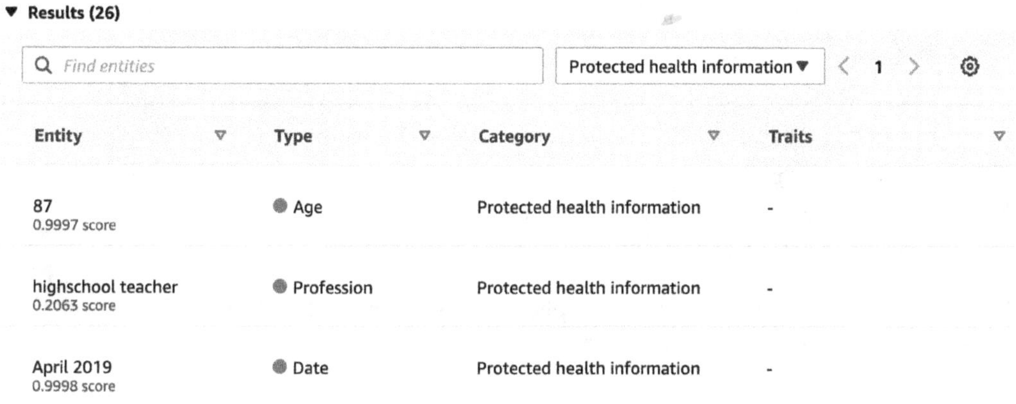

▼ **Results (26)**

| Q Find entities | | | Protected health information ▼ | | ⟨ 1 ⟩ | ⚙ |

Entity	▽	Type	▽	Category	▽	Traits	▽
87 0.9997 score		● Age		Protected health information		-	
highschool teacher 0.2063 score		● Profession		Protected health information		-	
April 2019 0.9998 score		● Date		Protected health information		-	

Figure 6.10 – PHI in Amazon Comprehend Medical

We just learned about PHI detection in Amazon Comprehend Medical on the AWS Console.

Now, let's dive into the implementation of code for redacting PII information:

1. Import the required libraries to call and parse the APIs:

```
import pandas as pd
import webbrowser, os
import json
import boto3
import re
import sagemaker
from sagemaker import get_execution_role
from sagemaker.s3 import S3Uploader, S3Downloader
import uuid
import time
import io
from io import BytesIO
import sys
from pprint import pprint
from IPython.display import Image, display
from PIL import Image as PImage, ImageDraw
```

2. Get the SageMaker default region, role, and bucket, as shown in this sample code:

```
region = boto3.Session().region_name
role = get_execution_role()
print(role)
bucket = sagemaker.Session().default_bucket()
prefix = "pii-detection-redaction"
bucket_path = "https://s3-{}.amazonaws.com/{}".
format(region, bucket)
```

3. We are using a sample employment application document for this example. Let's check the sample image with the following code:

```
# Document
documentName = "employmentapp.png"
display(Image(filename=documentName))
```

Here, you can see the sample employment application document:

Employment Application

Applicant Information

Full Name: Jane Doe

Phone Number: 555-0100|

Home Address: 123 Any Street, Any Town, USA

Mailing Address: same as home address

Previous Employment History				
Start Date	End Date	Employer Name	Position Held	Reason for leaving
1/15/2009	6/30/2011	Any Company	Assistant Baker	Family relocated
7/1/2011	8/10/2013	Best Corp.	Baker	Better opportunity
8/15/2013	present	Example Corp.	Head Baker	N/A, current employer

Figure 6.11 – An employment application

4. The first thing to do is extract all the data from the document. To process this single-page document, we are using Tetxract's synchronous `detect_document_text` API, which takes image bytes. Thus, we are collecting raw bytes from the employment application first. Subsequently, we get the `response` object from the Textract API:

```
with open(documentName, 'rb') as file:
        img_test = file.read()
        bytes_test = bytearray(img_test)
```

```
        print('Image loaded', documentName)
    # process using image bytes
response = textract.detect_document_
text(Document={'Bytes': bytes_test})
```

5. In the following code sample, we are collecting all the LINES extracted by the Textract API:

```
#Extract key values
from trp import Document
doc = Document(response)
page_string = ''
for page in doc.pages:
    # Print lines and words
        for line in page.lines:
            page_string += str(line.text)
print(page_string)
```

6. Now, check the screenshot of the LINES of the sample employment application document extracted by Amazon Textract:

Employment ApplicationApplicant InformationFull Name: Jane DoePhone Number: 555-0100Home Address: 123 Any Street, A
ny Town, USAMailing Address: same as home addressPrevious Employment HistoryStart DateEnd DateEmployer NamePosition
HeldReason for leaving1/15/20096/30/2011Any CompanyAssistant BakerFamily relocated7/1/20118/10/2013Best Corp.BakerB
etter opportunity8/15/2013presentExample Corp.Head BakerN/A, currentemployer

Figure 6.12 – The LINES extracted from the employment application

7. Now, let's start the process of identifying PII information in the text. To process it, first, we are writing the content to a .txt file, pii_data.txt, with the following sample code:

```
# Lets get the  data into a text file
text_filename = 'pii_data.txt'
doc = Document(response)
with open(text_filename, 'w', encoding='utf-8') as f:
    for page in doc.pages:
    # Print lines and words
        page_string = ''
        for line in page.lines:
            #print((line.text))
            page_string += str(line.text)
        #print(page_string)
        f.writelines(page_string + "\n")
```

8. Let's write the local file to the Amazon S3 bucket now. We are using the SageMaker default bucket for our execution, creating an Amazon S3 `boto3` client, and uploading the local file with PII data to the Amazon S3 bucket by calling the `upload_file()` API. We are also storing S3 URIs for the follow-up Amazon Comprehend API call:

```
import sagemaker
# Load the documents locally for later analysis
with open(text_filename, "r") as fi:
    raw_texts = [line.strip() for line in fi.readlines()]
data_bucket = sagemaker.Session().default_bucket()
s3 = boto3.resource('s3')
s3.Bucket(data_bucket).upload_file("pii_data.txt",
"identified-data/pii_data.txt")
#print(sagemaker.Session().default_bucket())
InputS3URI = "s3://"+data_bucket+"/identified-data/pii_
data.txt"
OutputS3URI = "s3://"+data_bucket+"/non_pii_data.txt"
print(InputS3URI)
```

9. We are using Comprehend's PII detection feature to identify PII information from the text. To do that, first, we are getting Comprehend's `boto3` client.

10. Then, we are calling Comprehend's `start_pii_detection_job` asynchronous API. This API takes the `InputS3` and `OutputS3` URIs created in the previous step:

```
comprehend = boto3.client(service_name='comprehend')
role = sagemaker.get_execution_role()
response = comprehend.start_pii_entities_detection_job(
    InputDataConfig={
        'S3Uri': InputS3URI,
        'InputFormat': 'ONE_DOC_PER_FILE'
    },
    OutputDataConfig={
        'S3Uri': OutputS3URI
    },
    Mode='ONLY_REDACTION',
    RedactionConfig={
        'PiiEntityTypes': [
            'ALL',
        ],
        'MaskMode': 'MASK',
```

```
            'MaskCharacter': '*'
        },
        DataAccessRoleArn = role,
        JobName="job1",
        LanguageCode='en',
    )
```

11. As we used Comprehend's asynchronous API, we have to wait for the job to finish. If you need to check the status of the job, you can use the following sample code:

```
# Get the job ID
events_job_id = response['JobId']
job = comprehend.describe_pii_entities_detection_
job(JobId=events_job_id)
print(job)
```

In the preceding code, we are detecting all the PII elements in the document, but you have the option to pick PII elements selectively. To check the list of all the PII elements supported by Amazon Comprehend, please check the *References* section.

Follow these steps to check the status of the PII job:

1. Go to **Amazon Comprehend** on the **AWS Console** and click **Launch Amazon Comprehend**:

Figure 6.13 – Amazon Comprehend in the AWS Console

2. Click **Analysis jobs** on the left-hand panel. Then, it will show all the executed analysis jobs, including the PII analysis job named job1 from our previous steps.

3. You can also check the status of the analysis job. Wait for the analysis job status to change to **Completed** before proceeding:

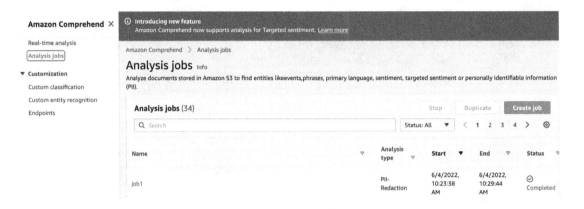

Figure 6.14 – Amazon Comprehend analysis jobs

4.   After the job status is **Completed**, click on the job. It shows the details of the job with output data location for redacted file, as shown in the sample screenshot here:

Figure 6.15 – Amazon Comprehend analysis job details

5.   Now, check the sample output file with its redactions:

```
Employment ApplicationApplicant InformationFull Name: ************ Number: *********** Address: ***
Any Street, Any Town, ********** Address: same as home addressPrevious Employment HistoryStart
DateEnd DateEmployer NamePosition HeldReason for leaving1******************** CompanyAssistant
BakerFamily relocated7******************* Corp.BakerBetter opportunity8*********************
Corp.Head BakerN/A, currentemployer
```

Figure 6.16 – The sample employment application with redactions

We just learned about how to programmatically implement sensitive information extraction and redaction with AWS AI services.

Redaction with structural elements

We are using an employment application document for this example:

1. We are gathering all the image bytes from this document in the following code:

    ```
    # Document
    documentName = "employmentapp.png"
    # Call Amazon Textract
    with open(documentName, 'rb') as document:
        imageBytes = bytearray(document.read())
    ```

2. We are calling the Textract `analyze_document` API to extract all key-value pairs from this document. We pass all the raw bytes extracted in the previous step to the API. We also pass `FeatureTypes` as `"FORMS"` to extract the key-value pairs from this document:

    ```
    # Call Amazon Textract
    response = textract.analyze_document(Document={'Bytes':
    imageBytes}, FeatureTypes=["FORMS"])
    #print(response)
    doc = Document(response)
    ```

3. Amazon Textract also gives geometrical information – for example, `boundingBox` – about the extracted element. We are leveraging the `boundingBox` feature of Amazon Textract to redact the `value` field from the document:

    ```
    # Redact document
    img = PImage.open(documentName)
    width, height = img.size
    if(doc.pages):
        page = doc.pages[0]
        for field in page.form.fields:
            if(field.key and field.value and "address" in
    field.key.text.lower()):
                print("Redacting => Key: {}, Value: {}".
    format(field.key.text, field.value.text))
                x1 = field.value.geometry.boundingBox.
    left*width
                y1 = field.value.geometry.boundingBox.
    ```

```
top*height-2
            x2 = x1 + (field.value.geometry.boundingBox.
width*width)+5
            y2 = y1 + (field.value.geometry.boundingBox.
height*height)+2
            draw = ImageDraw.Draw(img)
            draw.rectangle([x1, y1, x2, y2],
fill="Black")
outputDocumentName = "redacted-{}".format(documentName)
img.save(outputDocumentName)
display(Image(filename=outputDocumentName))
```

4. For our code implementation, we want to redact the **Home Address** and **Mailing Address** values. Let's print the value of these key fields as shown here.

We are also displaying the redacted image file as shown:

```
Redacting => Key: Home Address:, Value: 123 Any Street, Any Town, USA
Redacting => Key: Mailing Address:, Value: same as home address
```

Employment Application

Applicant Information

Full Name: Jane Doe

Phone Number: 555-0100

Home Address: ██████████████████

Mailing Address: ████████████████

Previous Employment History				
Start Date	**End Date**	**Employer Name**	**Position Held**	**Reason for leaving**
1/15/2009	6/30/2011	Any Company	Assistant Baker	Family relocated
7/1/2011	8/10/2013	Best Corp.	Baker	Better opportunity
8/15/2013	present	Example Corp.	Head Baker	N/A, current employer

Figure 6.17 – The employment application with redactions

We just learned about how to programmatically implement sensitive information extraction and redaction with AWS AI services, leveraging the structural element of the document. We can leverage the detection of sensitive PII or PHI entities from input documents and can redact and store them securely. Some additional security or privacy recommendations are to think about the security of your documents at rest. We have considered the redaction of input and output here. I recommend using encryption at rest for all your documents in the storage layer. AWS AI services store data for AI service improvement by default – if your company's policy or regulations don't permit this, you can use the AWS opt-out policy. Another thing to consider is security in transit. For security in transit, AWS AI services support the HTTPS protocol. Also, AWS AI services offer VPC endpoints to keep traffic within the AWS network. I recommend checking the *References* section for the details.

Learning about the document review process with human-in-the-loop

We have already discussed the IDP Extraction and Enrichment stages for document processing. But at times, the extraction ML models may not give you accurate information. The confidence score might be low in terms of your business needs. To meet your business needs and the threshold of confidence score, you can involve a human workforce to only review those elements. Having a human being review some or certain fields of your documents in an automated way for higher accuracy can also be a part of the post-processing phase in IDP. You can leverage human review not only for lower accuracy results but also for any critical field in your document.

Validation without automation

Without automation, all of the documents need to be processed with human assistance. This is more expensive and time-consuming. *Figure 6.18* shows the flow of the manual review process without automation:

Figure 6.18 – Validation without automation

Without automation, all elements require human review, which gets expensive.

Validation with automation

With IDP services, you can now automate document processing by including a human review only when it is required. Most often, 80% to 90% of the time, document processing can go straight through without human intervention using IDP services. You can set up a human review to only check for lower threshold elements extracted from documents. Moreover, you can send a random sample or only certain fields as required by your use case for human review. This will limit human review to certain scenarios, reducing time and costs. In *Figure 6.19*, we are setting a confidence threshold, and as per the confidence threshold, routing documents for human review when needed:

AFTER

Limit human review of documentation based on confidence scores < target to reduce total cost of document processing

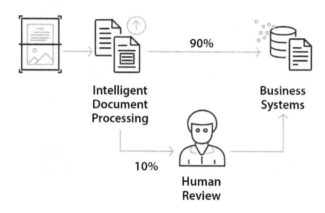

Figure 6.19 – Validation with automation

With automation, only part of the process requires human review, helping in reducing the cost and time to process documents.

Amazon Augmented AI (Amazon A2I) for the human review process

Amazon A2I is an ML service that makes it easy to build an IDP pipeline with human review. Amazon A2I takes care of building the human review system and managing the human review workforce.

In the following figure, we integrated Amazon A2I into our IDP pipeline to deliver the Review and Validation step of IDP:

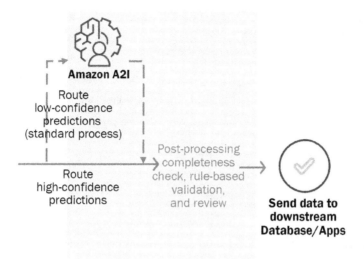

Verification and Human Review

Figure 6.20 – Human Review and Verification in IDP

Now, let's look at the following code sample to automate the human review process in IDP.

We will look into Amazon A2I for automating the document review process. The human review process broadly consists of five steps:

1. Create the human review workflow. To follow the steps, it is recommended to check A2I's example of a default Textract A2I FORM template in the following reference: https://docs.aws.amazon.com/sagemaker/latest/dg/a2i-create-flow-definition.html.

2. Set up A2I for the document review process. We will use the following sample code to automate this.

    I.   Create a `humanLoopConfig`. This takes Arn of Flow definition:

```
uniqueId = str(uuid.uuid4())
human_loop_unique_id = uniqueId + '1'
print(human_loop_unique_id)
humanLoopConfig = {
    'FlowDefinitionArn':"<FLOW_DEFINITION_ARN>",
    'HumanLoopName':human_loop_unique_id,
    'DataAttributes': { 'ContentClassifiers': [
'FreeOfPersonallyIdentifiableInformation' ]}
}
```

II.   Amazon Textract is seamlessly integrated with Amazon A2I for its synchronous `analyze_document` API. While calling the Textract `analyze_document` API, we pass in the `human_loop_config` created in the preceding step:

```
def analyze_document_with_a2i(document_name, bucket):
    response = textract.analyze_document(
        Document={'S3Object': {'Bucket': bucket, 'Name':
document_name}},
        FeatureTypes=["TABLES", "FORMS"],
        HumanLoopConfig=humanLoopConfig
    )
    return response
```

III.   You may not want to send all the information for human review, but can customize the business rules to send only required information for human review. For example, you may want to send key-value fields which are lower confidence threshold or may want to send ca critical field such as ID number from IRS document for additional human review. If you want to learn how to customize business rules for human review please check the reference below.

IV.   Once completed, you can check the consolidated output in your given Amazon S3 bucket.. I would recommend to check A2I output in the reference.

Note

For a full code sample implementation of the IDP pipeline with Amazon A2I, please check this link: `https://github.com/aws-samples/amazon-a2i-sample-jupyter-notebooks`. We are covering Human review process with Amazon A2I, but business can use their existing human review solution for further post processing of documents.

Summary

In this chapter, we discussed the core features of Amazon Comprehend, PII detection and redaction, and Amazon Comprehend Medical's PHI detection feature. We also discussed the Review and Validation stage of the IDP pipeline and why it is critical for accurate IDP. We also discussed how to leverage Amazon Textract to extract text from any type of document and then pass it to Amazon Comprehend (Medical) for PII or PHI information detection and redaction. This helps to build a document processing pipeline to handle sensitive information.

We then reviewed the need for human review. We also discussed Amazon A2I and its core features for including human beings in the review of more critical field elements in documents, or ones with lower accuracy. This automation helps build cost-effective document processing with time acceleration.

In the next chapter, we will discuss how to build a data lake for health information and how IDP can be integrated with Amazon HealthLake to deliver a data lake, whether structured or unstructured.

References

- *PII data*: https://www.dol.gov/general/ppii

- *PII entities supported with Amazon Comprehend* – https://docs.aws.amazon.com/comprehend/latest/dg/how-pii.html

- *AWS AI Services opt-out policy*: https://docs.aws.amazon.com/organizations/latest/userguide/orgs_manage_policies_ai-opt-out.html

- *VPC endpoints*: https://docs.aws.amazon.com/textract/latest/dg/vpc-interface-endpoints.html

- *A2I Template*: https://docs.aws.amazon.com/sagemaker/latest/dg/a2i-create-flow-definition.html

- *A2I Customize Business Rules*: https://github.com/aws-samples/aws-ai-intelligent-document-processing/blob/main/04.01-idp-a2i-with-custom-rules.ipynb

- *A2I reference output*: https://github.com/aws-samples/aws-ai-intelligent-document-processing/blob/main/04.01-idp-a2i-with-custom-rules.ipynb [Check A2I generated JSON]

7

Accurate Extraction, and Health Insights with Amazon HealthLake

In the previous chapter, we examined the challenges in the review and verification stage of the **Intelligent Document Processing (IDP)** pipeline. We looked at how we can leverage business rules with serverless architecture to automate validation, such as checking the completeness of a document. We also discussed how to leverage and automate human review with AWS AI services and discussed completeness checks with post-processing logic. We learned how to use the Amazon Comprehend PII and Amazon Comprehend Medical PHI detection APIs to handle sensitive data. We will now change gear and dive into document processing with clinical health data extraction and insights with Amazon HealthLake. We will navigate through the following sections in this chapter:

- Introducing **Fast Healthcare Interoperability Resources (FHIR)**
- Using Amazon HealthLake as a health data store
- Handling documents with an FHIR data store

Technical requirements

For this chapter, you will need access to an AWS account. Before getting started, we recommend that you create an AWS account by referring to the AWS account setup and Jupyter notebook creation steps mentioned in the *Technical requirements* section in *Chapter 2, Document Capture and Categorization*

You can find Chapter-7 code sample in GitHub: `https://github.com/PacktPublishing/Intelligent-Document-Processing-with-AWS-AI-ML-/tree/main/chapter-7`. Also recommend to check availability of AI service in your AWS regions before using it.

Introducing Fast Healthcare Interoperability Resources (FHIR)

Despite the widespread usage and adoption of **Electronic Health Records** (**EHRs**), one-third of providers, payers, and care teams struggle to exchange healthcare data. A patient may go through multiple doctor visits, routine check-ups, and lab tests over time, and we should treat all this data as essential. But this data is siloed and doesn't provide a central patient view, which thus makes the goal to catalog the entire patient's medical journey.

Healthcare data interoperability is a step toward combining health data across various disparate systems and sites to help healthcare professionals to spend more time with their patients rather than performing healthcare data collection.

Although healthcare organizations define interoperability standards, it is not enough. While data standards have been available in the past, they have not been sufficient to achieve full interoperability. For interoperability to be successful, healthcare organizations need a more granular focus that encompasses syntactic interoperability, where a common standard and language are used to interpret the data.

The **Health Level Seven**® (**HL7**®) **Fast Healthcare Interoperability Resources** (**FHIR**®) standard defines how healthcare information can be exchanged between different systems. Data can be stored in different formats and on disparate systems. One of the main requirements of FHIR is to follow secure and confidential data exchange methods. The standards development organization HL7® uses a collaborative approach to develop and upgrade FHIR. FHIR offers extended operations with RESTful APIs to perform actions. You can use some of the common API operations, such as **CREATE, READ, UPDATE, and DELETE** (**CRUD**), to interact and perform actions on health data resources in the repository. We will be using FHIR CRUD operations on our healthcare resources, so we recommend getting familiar with FHIR operations.

Using Amazon HealthLake as a health data store

Amazon HealthLake is an end-to-end HIPAA-eligible machine learning service that provides health, healthcare, and life sciences customers with a complete view of their patients, with analytic and query functionality. What do we mean by HIPAA eligible? An HIPAA-eligible AI service is one that can be configured to meet HIPAA compliance requirements. For example, the service offers different encryption mechanisms to support security at rest, but it is the responsibility of a person to configure the required type of encryption mechanism as per their compliance requirements. *Figure 7.1* is a high-level diagram of Amazon HealthLake being fed both structured and unstructured health data. Health data is enriched and normalized for further analytics, search, or machine learning use cases.

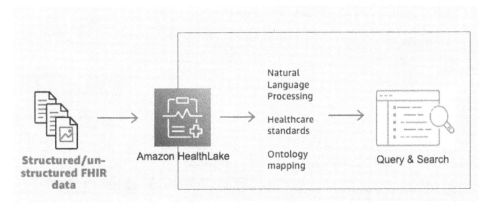

Figure 7.1 – Amazon HealthLake

You can input any health data, such as medical reports, doctor's notes, and lab reports, in FHIR format to Amazon HealthLake. Amazon HealthLake stores, transforms, and normalizes this input, and derives meaningful health insights by leveraging machine learning to learn from raw health data. The secure HIPAA-eligible infrastructure of Amazon HealthLake further enriches the data with medical insights and ontology linking for queries and search operations. Often, customers extend the transformed and indexed data with Amazon SageMaker for model development or Amazon QuickSight and additional analytics services for visualization.

Now let's take a look at Amazon HealthLake's features in the AWS console:

1. Go to **Amazon HealthLake** in the AWS console and click on **View Data Stores**:

Figure 7.2 – Amazon HealthLake on AWS Console

2.   We need to create a data store. Click on **Create Data Store**:

Figure 7.3 – Amazon HealthLake – Create Data Store

3.   Provide a name for your data store and leave the **Format** as **FHIR R4**. If you want, you can load sample data to explore the functionality of Amazon HealthLake. It will load your data store with **Synthea** synthetic data.

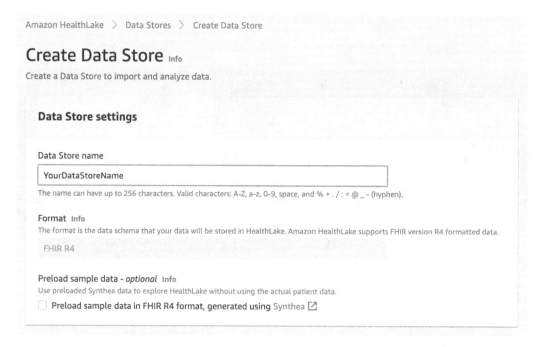

Figure 7.4 – Create Data Store settings

4. Amazon HealthLake offers encryption at rest with Amazon KMS, but for our experiment, we will use AWS owned keys:

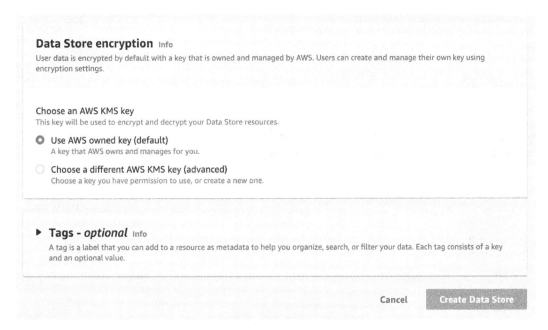

Figure 7.5 – Data store encryption settings

5. You can check that the data store has been created successfully as follows:

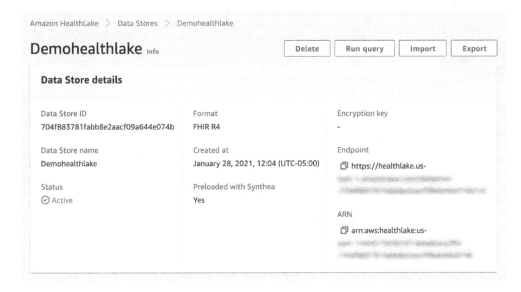

Figure 7.6 – Amazon HealthLake data store

We have been through the steps to create a data store in Amazon HealthLake. Now let's see how to perform FHIR operations against this data.

FHIR operations with Amazon HealthLake

Let's run some operations on our Amazon HealthLake data store. Amazon HealthLake's console has a query interface for quick testing. Let's run some queries.

PUT operation

We are planning to PUT a patient FHIR resource on Amazon HealthLake data store. We will leverage the sample patient FHIR data from Synthea for testing with the following instructions:

1. Click on **Run query**, as shown in the preceding figure.

2. In **Query settings**, select **Create** for **Query type**, and select **Patient** for **Resource type**. If you click on **Info** next to **Request body**, the sample input data opens up in the right-hand panel. Click on **Apply Patient example**, and this will fill the **Request examples** section with sample data.

Figure 7.7 – Amazon HealthLake - Create Patient Resource

3. Then click on **Run query**:

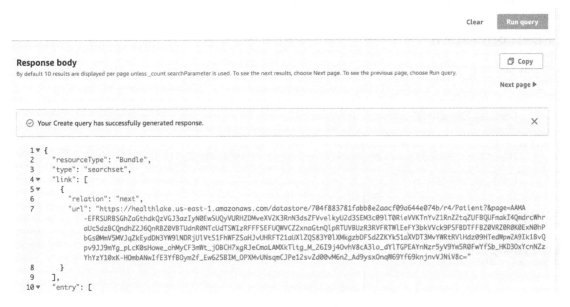

Figure 7.8 – Amazon HealthLake query – Run query

4. You can look at the response, as shown in the following figure. Amazon HealthLake will create and store FHIR patient resources in a scalable data store. Make a note of the `"id"` from the Patient FHIR resource.

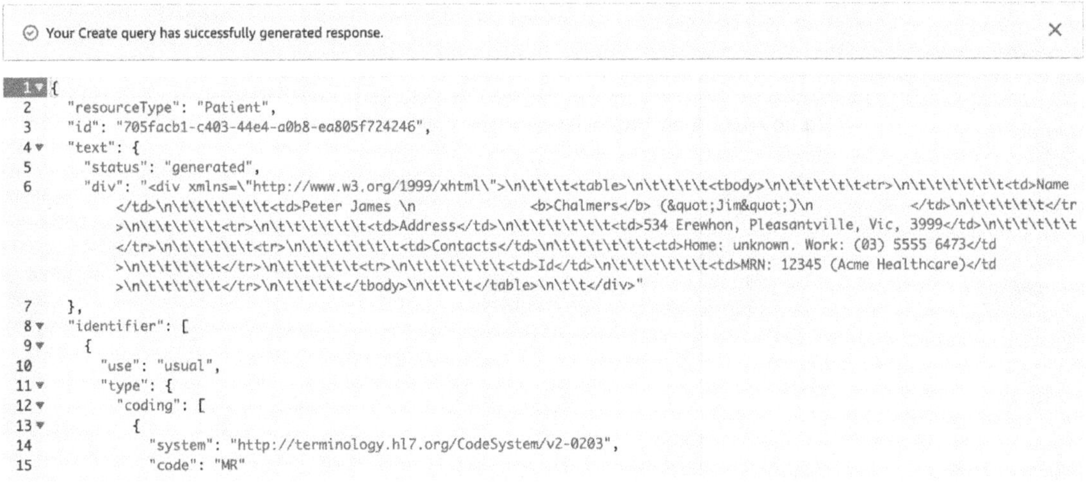

Figure 7.9 – Amazon HealthLake query – response

We just looked at an example of running an FHIR PUT operation on Amazon HealthLake. Now let's see how to run an FHIR READ operation to read FHIR data from our data store.

READ operation

We are planning to do a READ operation on our patient's FHIR resource on our Amazon HealthLake data store. We will leverage the Synthea sample patient FHIR data with the following instructions:

1. Click on **Run query**, as shown in *Figure 7.10*.

2. In **Query settings**, select **Read** as **Query type**, **Patient** as **Resource type**, and for the **Resource ID**, provide the ID copied in the *PUT operation* section in *step 4*.

3. Click on **Run query**:

Figure 7.10 – Amazon HealthLake query – Read request

The patient resource is shown in the following figure:

Response body   ⎘ Copy

By default 10 results are displayed per page unless _count searchParameter is used. To see the next results, choose Next page. To see the previous page, choose Run query.

Next page ▶

⊘ Your Read query has successfully generated response.   ✕

```
 1 ▾ {
 2    "resourceType": "Patient",
 3    "id": "705facb1-c403-44e4-a0b8-ea805f724246",
 4 ▾  "text": {
 5      "status": "generated",
 6      "div": "<div xmlns=\"http://www.w3.org/1999/xhtml\">\n\t\t\t<table>\n\t\t\t\t<tbody>\n\t\t\t\t\t<tr>\n\t\t\t\t\t\t<td>Name
         </td>\n\t\t\t\t\t\t<td>Peter James \n              <b>Chalmers</b> (&quot;Jim&quot;)\n              </td>\n\t\t\t\t\t</tr
         >\n\t\t\t\t\t<tr>\n\t\t\t\t\t\t<td>Address</td>\n\t\t\t\t\t\t<td>534 Erewhon, Pleasantville, Vic, 3999</td>\n\t\t\t\t\t
         </tr>\n\t\t\t\t\t<tr>\n\t\t\t\t\t\t<td>Contacts</td>\n\t\t\t\t\t\t<td>Home: unknown. Work: (03) 5555 6473</td
         >\n\t\t\t\t\t</tr>\n\t\t\t\t\t<tr>\n\t\t\t\t\t\t<td>Id</td>\n\t\t\t\t\t\t<td>MRN: 12345 (Acme Healthcare)</td
         >\n\t\t\t\t\t</tr>\n\t\t\t\t</tbody>\n\t\t\t</table>\n\t\t</div>"
 7    },
 8 ▾  "identifier": [
 9 ▾    {
10        "use": "usual",
11 ▾      "type": {
12 ▾        "coding": [
13 ▾          {
14              "system": "http://terminology.hl7.org/CodeSystem/v2-0203",
15              "code": "MR"
16            }
```

Figure 7.11 – Amazon HealthLake read response

We have been through the steps to upload a sample patient FHIR resource to our Amazon HealthLake data store, and then we used the READ API to read the uploaded patient resource from the FHIR data store. Amazon HealthLake supports **Create, Read, Update, Delete (CRUD)** operations.

Transform

Amazon HealthLake has **natural language models** to transform raw medical text with medical NLP, such as extracting an entity's medical condition, medication, dosage, traits, and PHI. We will get more details on that in the following examples.

We will now look at Amazon HealthLake operations that leverage APIs programmatically.

For full code-walkthrough follow the steps in the Notebook: `https://github.com/PacktPublishing/Intelligent-Document-Processing-with-AWS-AI-ML-/blob/main/chapter-7/healthlake-07.ipynb`

HealthLake PUT request

We have created a data store, and now we will use the FHIR API to PUT FHIR resources such as patient data in the data store:

1. Import the libraries required to run an FHIR PUT request on Amazon HealthLake:

   ```
   import boto3
   import requests
   import json
   ```

2. Then we create a boto3 client for Amazon HealthLake:

   ```
   client = boto3.client('healthlake')
   ```

3. Get FHIR HealthLake `datastore` endpoint and patient resource, and store them in `data_store_endpoint` and `resource_path` variables. We will use put_file. json (`https://github.com/PacktPublishing/Intelligent-Document-Processing-with-AWS-AI-ML-/blob/main/chapter-7/put_file.json`) to create patient FHIR resource.

   ```
   # Parse the input arguments
   data_store_endpoint = "<YOUR_DATASTORE_ENDPOINT>"
   resource_path = "Patient/3cedeb4b-a9b8-41d2-aef2-
   de4c0413ec95" //replace this with your unique patient ID
   request_body_file = "put_file.json"
   region = "us-east-1"
   # Frame the resource endpoint
   resource_endpoint = data_store_endpoint+resource_path
   ```

4. We now need authentication for our FHIR API. We are using the `requests_auth_aws_sigv4` library for a SigV4 signature:

   ```
   !pip install requests-auth-aws-sigv4
   from requests_auth_aws_sigv4 import AWSSigV4
   region = "us-east-1"
   session = boto3.session.Session(region_name=region)
   # Frame authorization
   auth = AWSSigV4("healthlake", session=session)
   ```

5. As mentioned in *step 3*, our patient resource JSON is defined in put_file.json. We need to read the JSON body and store the data in json_data.

6. Then we call the FHIR endpoint with SigV4 authentication. We are calling a PUT request and passing json_data with authentication to the put method:

```
# Read the request body from input file
with open(request_body_file) as json_body:
    json_data = json.load(json_body)
# Calling data store FHIR endpoint using SigV4 auth
r = requests.put(resource_endpoint, json=json_data,
auth=auth)
print(r.json())
```

You can see the patient resource result in the following figure:

```
Requirement already satisfied: requests-auth-aws-sigv4 in /home/ec2-user/anaconda3/envs/python3/lib/python3.6/site-
packages (0.7)
Requirement already satisfied: requests in /home/ec2-user/anaconda3/envs/python3/lib/python3.6/site-packages (from
requests-auth-aws-sigv4) (2.26.0)
Requirement already satisfied: charset-normalizer~=2.0.0 in /home/ec2-user/anaconda3/envs/python3/lib/python3.6/sit
e-packages (from requests->requests-auth-aws-sigv4) (2.0.9)
Requirement already satisfied: certifi>=2017.4.17 in /home/ec2-user/anaconda3/envs/python3/lib/python3.6/site-packa
ges (from requests->requests-auth-aws-sigv4) (2021.5.30)
Requirement already satisfied: urllib3<1.27,>=1.21.1 in /home/ec2-user/anaconda3/envs/python3/lib/python3.6/site-pa
ckages (from requests->requests-auth-aws-sigv4) (1.26.7)
Requirement already satisfied: idna<4,>=2.5 in /home/ec2-user/anaconda3/envs/python3/lib/python3.6/site-packages (f
rom requests->requests-auth-aws-sigv4) (3.1)
{'id': '3cedeb4b-a9b8-41d2-aef2-de4c0413ec95', 'resourceType': 'Patient', 'active': True, 'name': [{'use': 'officia
l', 'family': 'Dee', 'given': ['Jane']}, {'use': 'usual', 'given': ['Jane']}], 'gender': 'female', 'birthDate': '19
66-09-01', 'meta': {'lastUpdated': '2022-05-31T18:58:46.602Z'}}
```

Figure 7.12 – Patient resource

We have learned how to perform PUT FHIR operations programmatically on Amazon HealthLake.

HealthLake GET request

We have created a data store and uploaded a patient FHIR resource. Now we will use the GET FHIR API (get(resource_endpoint, auth)) to get the patient FHIR resource created in the previous section:

1. Get FHIR HealthLake datastore endpoint and patient resource, and store them in data_store_endpoint and resource_path variables. We are trying to get the same patient resource that we created in *step 3* in the previous section:

```
data_store_endpoint = "<YOUR_DATASTORE_ENDPOINT>"
resource_path = "Patient/3cedeb4b-a9b8-41d2-aef2-
de4c0413ec95" //replace this with your unique patient ID
region = "us-east-1"
# Frame the resource endpoint
resource_endpoint = data_store_endpoint+resource_path
```

2. We need SigV4 authentication to call the FHIR API:

```
# Frame authorization
auth = AWSSigV4("healthlake", session=session)
```

3. Now we call the GET FHIR API and pass in the resource endpoint created in *step 1* and the authentication token created in *step 2*:

```
# Calling data store FHIR endpoint using SigV4 auth
r = requests.get(resource_endpoint, auth=auth)
print(r.json())
```

4. You can see the output of the GET request as follows:

{'id': '3cedeb4b-a9b8-41d2-aef2-de4c0413ec95', 'resourceType': 'Patient', 'active': True, 'name': [{'use': 'officia
l', 'family': 'Dee', 'given': ['Jane']}, {'use': 'usual', 'given': ['Jane']}], 'gender': 'female', 'birthDate': '19
66-09-01', 'meta': {'lastUpdated': '2022-05-31T18:58:46.602Z'}}

Figure 7.13 – Patient Resource – output

We have learned how to perform GET FHIR operations programmatically on Amazon HealthLake.

HealthLake delete request

We have created a data store and uploaded a patient FHIR resource. Now we will use the FHIR delete API to delete the patient FHIR resource:

1. Get FHIR HealthLake datastore endpoint and patient resource, and store them in `data_store_endpoint` and `resource_path` variables. We are trying to get the same patient resource that we created earlier:

```
data_store_endpoint = "<YOUR_DATASTORE_ENDPOINT>"
resource_path = "Patient/3cedeb4b-a9b8-41d2-aef2-
de4c0413ec95" //replace this with your unique patient ID
region = "us-east-1"
#Frame the resource endpoint
resource_endpoint = data_store_endpoint+resource_path
```

2. We need SigV4 authentication to call the FHIR API:

```
# Frame authorization
auth = AWSSigV4("healthlake", session=session)
```

3. Now we call the delete FHIR API and pass in the resource endpoint created in *step 1* and the authentication token created in *step 2*:

```
# Calling data store FHIR endpoint using SigV4 auth
r = requests.delete(resource_endpoint, auth=auth)
```

```
# HTTP Response code should be 204 once the resource is
successfully deleted
print(r)
```

4. We get a **Response 204** to show that the FHIR resource was successfully deleted, as shown in the following screenshot:

<Response [204]>

Figure 7.14 – Patient Resource – delete response

Now we understand the features of Amazon HealthLake and how to use FHIR APIs to perform operations on an FHIR data store.

Handling documents with an FHIR data store

Now let's see what happens when our input for HealthLake is not in FHIR format, but is a document instead. Amazon HealthLake currently only supports data in FHIR format. What if we need to process document-based health data along with FHIR data? Can we still create a centralized scalable FHIR data store for health data?

And the answer is yes. We can use Amazon Textract to get raw text from the document and convert it to an FHIR resource (a DocumentReference resource). This DocumentReference FHIR resource can then be input into Amazon HealthLake with additional FHIR resources.

This is a three-step process:

1. Extracting data from the document with Amazon Textract
2. Creating a DocumentReference FHIR resource from the extracted Textract response
3. Ingesting the DocumentReference FHIR resource to Amazon HealthLake

You can see the architecture in the following figure:

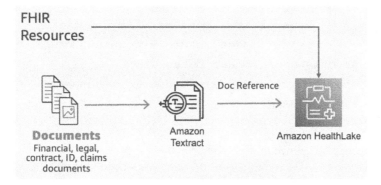

Figure 7.15 – Document processing with Amazon HealthLake

Now let's see the implementation of this architecture:

1. We are using a raw doctor's note for our program. This is a scanned image document. Let's display the doctor's note document:

```
# Document
documentName = "doctornotes1.png"
display(Image(filename=documentName))
```

2. The image of the doctor's note is as follows:

> The patient is an 86-year-old female admitted for evaluation of abdominal pain and bloody stools. The patient has colitis and also diverticulitis, undergoing treatment. During the hospitalization, the patient complains of shortness of breath, which is worsening. The patient underwent an echocardiogram, which shows severe mitral regurgitation and also large pleural effusion. This consultation is for further evaluation in this regard. As per the patient, she is an 86-year-old female, has limited activity level. She has been having shortness of breath for many years. She also was told that she has a heart murmur, which was not followed through on a regular basis.

Figure 7.16 – Sample doctor's note

3. Now we will call Amazon Textract to extract data from the doctor's note. We create a Textract client. Then we call the Textract API to extract all the elements from the document. We define the Textract call in a method called `calltextract()`:

```
# process using image bytes
def calltextract(documentName):
    client = boto3.client(service_name='textract',
        region_name= 'us-east-1',
        endpoint_url='https://textract.us-east-1.
amazonaws.com')
    with open(documentName, 'rb') as file:
            img_test = file.read()
            bytes_test = bytearray(img_test)
            print('Image loaded', documentName)
    # process using image bytes
    response = client.analyze_document(Document={'Bytes':
bytes_test}, FeatureTypes=['FORMS'])
    return response
```

4. Now call the method defined in *step 3* and print the response:

```
response= calltextract(documentName)
print(response)
```

5. You can see the response in the next figure:

Image loaded doctornotes1.png
{'DocumentMetadata': {'Pages': 1}, 'Blocks': [{'BlockType': 'PAGE', 'Geometry': {'BoundingBox': {'Width': 1.0, 'Hei
ght': 1.0, 'Left': 0.0, 'Top': 0.0}, 'Polygon': [{'X': 4.247091688087781e-17, 'Y': 0.0}, {'X': 1.0, 'Y': 3.53126315
1746437e-16}, {'X': 1.0, 'Y': 1.0}, {'X': 0.0, 'Y': 1.0}]}, 'Id': 'c29dce4c-09a9-4d7b-84fd-e7eb90b66718', 'Relation
ships': [{'Type': 'CHILD', 'Ids': ['960e51b0-f556-4205-92dd-f01848c9bde5', 'a67f18bd-9534-421b-a3ca-2db71a610a77',
'a9a3d1d5-e59f-4c7f-8331-7dd6952c06f2', '7672eb9b-889c-4205-b0e7-33b1d1df31a4', '1fe55fd2-8045-4b8f-afcc-68e20edef8
6b', '8aea2404-cfdf-4d15-81ee-35a80bcca1fc', '74f221a4-2091-456b-9e7a-1c960d904315', '2be35b49-7b4f-4bb9-8a9a-6b147
09c042e', '6864ffb7-8301-41f4-a0a2-d4cfb7805169', '93c2c23e-e468-4811-877d-e6ed331e642f', '4ed36cab-371d-4f10-b3df-
0f302b1a47ae', '53636e3b-43ca-4d23-8eb7-fd036252d87e', '320fb687-0af0-4a50-862e-63ffcffb988f', '997f6ec6-0740-4007-
8ccf-c02f1ce45add', '2b25902c-4de5-440e-ba73-643be08a2610', '35df383a-bdaa-417d-b341-d59fd25293b8', '75c24d2b-7f28-
4523-aedc-de191b5878a0']}]}, {'BlockType': 'LINE', 'Confidence': 99.94792175292969, 'Text': 'The patient is an 86-y
ear-old female admitted for evaluation of abdominal pain and', 'Geometry': {'BoundingBox': {'Width': 0.811660647392
273, 'Height': 0.06564337015151978, 'Left': 0.03203154355287552, 'Top': 0.1808532327413559}, 'Polygon': [{'X': 0.03

Figure 7.17 – Textract response

6. Print all the lines extracted from the doctor's note and store them in the `text` variable. Also, let's print all the lines from the document:

```
# Print detected text
text = ""
for item in response["Blocks"]:
    if item["BlockType"] == "LINE":
        text = text + " " + item["Text"]


print(text)
```

7. This will print the lines extracted by Amazon Textract from the raw doctor's note:

The patient is an 86-year-old female admitted for evaluation of abdominal pain and bloody stools. The patient has
colitis and also diverticulitis, undergoing treatment. During the hospitalization, the patient complains of shortne
ss of breath, which is worsening. The patient underwent an echocardiogram, which shows severe mitral regurgitation
and also large pleural effusion. This consultation is for further evaluation in this regard. As per the patient, sh
e is an 86-year-old female, has limited activity level. She has been having shortness of breath for many years. She
also was told that she has a heart murmur, which was not followed through on a regular basis.

Figure 7.18 – Raw doctor's note

8. Next, we create a DocumentReference FHIR resource. The API requires a `base64` encoded string. So, we are using the `base64` library to encode the string:

```
#convert to base64
import base64
sample_string_bytes = text.encode("ascii")


base64_bytes = base64.b64encode(sample_string_bytes)
```

```
print(base64_bytes)
base64_string = base64_bytes.decode("ascii")

print(f"Encoded string: {base64_string}")
```

9. Let's look at the encoded string:

b'IFRoZSBwYXRpZW50IGlzIGFuIDg2LXllYXItb2xkIGZlbWFsZSBhZG1pdHRlZCBmb3IgZXZhbHVhdGlvbiBvZiBhYmRvbWluYWwgcGFpbBhbmQgYmxvb2R5IHN0b29scy4gVGhlIHBhdGllbnQgaGFzIGNvbGl0aXMgYW5kIGFsc28gZGl2ZXXJ0aWN1bGl0aXMsIHVuZGVyZ29pbmcgdHJlYXRtZW50LiBEdXJpbmcgdGhlIGhvc3BpdGFsaXphdGlvbiwgdGhlIHBhdGllbnQgY29tcGxhaW5zIG9mIHNob3J0bmVzcyBvZiBicmVhdGggGgIHdoaWNoIGlzIHdvcnNlbmluZy4gVGhlIHBhdGllbnQgdW5kZXJ3ZW50IGFuIGVjaG9jYXJkaW9ncmFtLCB3aGljaCBzaG93cyBzXZXZlcmUgbWl0cmFsIHJlZ3VyZ2l0YXRpb24gYW5kIGFsc28gbGVmdCUgdmVudHJpY3VsYXIgaHlwZXJ0cm9waHkuIFRoZSBwYXRpZW50LCBzaGGUgaXMgYW4gODlteWVhci1vbGQgZmVtYWxlLCBoYXMgbGltaXRlZCBhY3Rpdml0eSBSZXlbC4gU2hlIGhhcyBiZWVuIGhhdmluZyBzaG9ydG5lc3Mgb2YgYnJlYXRoIGZvciBtYW55IHllYXJzIiBiaTaGUgYWxzbyB3YXMgdG9sZCB0aGF0IHNoZSBoYXMgYSBoZWFydCBtdXJtdXIuIHdoaWNoIHdhcyBub3QgZm9sbG93ZWQgdGhyb3VnaCBpbiHJlZVsYXIgYmFzaXMu'
Encoded string: IFRoZSBwYXRpZW50IGlzIGFuIDg2LXllYXItb2xkIGZlbWFsZSBhZG1pdHRlZCBmb3IgZXZhbHVhdGlvbiBvZiBhYmRvbWluYWwgcGFpbBhbmQgYmxvb2R5IHN0b29scy4gVGhlIHBhdGllbnQgaGFzIGNvbGl0aXMgYW5kIGFsc28gZGl2ZXXJ0aWN1bGl0aXMsIHVuZGVyZ29pbmcgdHJlYXRtZW50LiBEdXJpbmcgdGhlIGhvc3BpdGFsaXphdGlvbiwgdGhlIHBhdGllbnQgY29tcGxhaW5zIG9mIHNob3J0bmVzcyBvZiBicmVhdGggGgIHdoaWNoIGlzIHdvcnNlbmluZy4gVGhlIHBhdGllbnQgdW5kZXJ3ZW50IGFuIGVjaG9jYXJkaW9ncmFtLCB3aGljaCBzaG93cyBzXZXZlcmUgbWl0cmFsIHJlZ3VyZ2l0YXRpb24gYW5kIGFsc28gbGVmdCUgdmVudHJpY3VsYXIgaHlwZXJ0cm9waHkuIFRoZSBwYXRpZW50LCBzaGGUgaXMgYW4gODlteWVhci1vbGQgZmVtYWxlLCBoYXMgbGltaXRlZCBhY3Rpdml0eSBSZXlbC4gU2hlIGhhcyBiZWVuIGhhdmluZyBzaG9ydG5lc3Mgb2YgYnJlYXRoIGZvciBtYW55IHllYXJzIiBiaTaGUgYWxzbyB3YXMgdG9sZCB0aGF0IHNoZSBoYXMgYSBoZWFydCBtdXJtdXIuIHdoaWNoIHdhcyBub3QgZm9sbG93ZWQgdGhyb3VnaCBpbiHJlZVsYXIgYmFzaXMu

Figure 7.19 – base64 encoded string

10. The next step is to create the DocumentReference FHIR resource with the encoded string created in *step 9*:

```
import json
json_docref = '{"resourceType":"DocumentReference",
"id":"id12345","meta":{"profile":["http://hl7.org/fhir
/us/core/StructureDefinition/us-core-
documentreference"]},
"identifier":[{"system":"urn:ietf:rfc:3986","value":"urn:
uuid:f11b9b48-ff6b-a62b-8425-037ea9c2d826"}],"status":
"current","type":{"coding":[{"system":"http://loinc.org",
"code":"34117-2","display":"History and physical note"},
{"system":"http://loinc.org","code":"51847-2","display":
"Evaluation+Plan note"}]},"category":[{"coding":[{"system"
:"http://hl7.org/fhir/us/core/CodeSystem/us-core-
documentreference-category","code":"clinical-note",
"display":"Clinical Note"}]}],"subject":{"reference":
"Patient/patientid12345"},"date":"2022-04-13T05:07:38.
982+00:00","author":[{"reference":"Practitioner?
identifier=http://hl7.org/fhir/sid/us-npi|9999990259",
"display":"Dr. Tandra334 Carter549"}],"custodian":
{"reference":"Organization?identifier=https://ocktank.
com/
synthetic/synthea|f7ae497d-8dc6-3721-9402-43b621a4e7d2",
```

```
"display":"PCP14023"},"content":[{"attachment":
{"contentType":"text/plain; charset=utf-8","data":
"IFRoZSBwYXRpZW50IGlzIGFuIDg2LXllYXItb2xkIGZlbWFsZSBhZG1p
dHRlZCBmb3IgZXZhHVhdGlvbiBvZiBhYmRvbWluYWwgcGFpbiBhbmQgY
mxvb25IHN0b29scy4gVGhlIHBhdGllbnQgaGFzIGNvbG9aXMgYW5kIG
Fsc28gZGl2ZXJ0aWN1bGl0aXMsIHVuZGVyZ29pbmcgdHJlYXRtZW50LiB
EdXJpbmcgdGhlIGhvc3BpdGFsaXphdGlvbiwgdGhlIHBhdGllbnQgY29t
cGxhaW5zIG9mIHNob3J0bmVzcyBvZiBicmVhdGgsIHdoaWNoIGlzIHdvc
nNlbmluZy4gVGhlIHBhdGllbnQgdW5kZXJ3ZW50IGFuIGVjaG9jYXJkaW
9ncmFtLCB3aGljaCBzaG93cyBzZXZlcmUgbWl0cmFsIHJlZ3VyZ2l0YXR
pb24gYW5kIGFsc28gbGFyZ2UgcGxldXJhbCBlZmZ1c2lvbi4gVGhpcyBj
b25zdWx0YXRpb24gaXMgZm9yIGZ1cnRoZXIgZXZhHVhdGlvbiBpbiB0a
GlzIHJlZ2FyZC4gQXMgcGVyIHRoZSBwYXRpZW50LCBzaGUgaXMgYW4gOD
YteWVhci1vbGQgZmVtYWxlLCBoYXMgbGltaXRlZCBhY3Rpdml0eSBsZXZ
lbC4gU2hlIGhhcyBiZWVuIGhhdmluZyBzaG9ydG5lc3Mgb2YgYnJlYXRo
IGZvciBtYW55IHllYXJzLiBTaGUgYWxzbyB3YXMgdG9sZCB0aGF0IHNoZ
SBoYXMgYSBoZWFydCBtdXJtdXIsIHdoaWNoIHdhcyBub3QgZm9sbG93ZW
QgdGhyb3VnaCBvbiBhIHJlZ3VsYXIgYmFzaXMu"},"format":
{"system":"http://ihe.net/fhir/ValueSet/IHE.FormatCode.
codesystem","code":"urn:ihe:iti:xds:2017:
mimeTypeSufficient","display":"mimeType Sufficient"}}],
"context":{"encounter":[{"reference":"Encounter/98
309c16-75dd-8780-4bab-2d2788f7f885"}],"period":{"s
tart":"2022-04-13T05:07:38+00:00","end":"2022-04-1
3T05:22:38+00:00"}}}'
print(json.dumps(json_docref))
```

11. You can see the printed JSON DocumentReference FHIR resource in the following screenshot. Also, we are manually storing the JSON DocumentReference FHIR in the doc_ref.json file:

"{\"resourceType\":\"DocumentReference\",\"id\":\"id12345\",\"meta\":{\"profile\":[\"http://hl7.org/fhir/us/core/St
ructureDefinition/us-core-documentreference\"]},\"identifier\":[{\"system\":\"urn:ietf:rfc:3986\",\"value\":\"urn:u
uid:f11b9b48-ff6b-a62b-8425-037ea9c2d826\"}],\"status\":\"current\",\"type\":{\"coding\":[{\"system\":\"http://loin
c.org\",\"code\":\"34117-2\",\"display\":\"History and physical note\"},{\"system\":\"http://loinc.org\",\"cod
e\":\"51847-2\",\"display\":\"Evaluation+Plan note\"}]},\"category\":[{\"coding\":[{\"system\":\"http://hl7.org/fhi
r/us/core/CodeSystem/us-core-documentreference-category\",\"code\":\"clinical-note\",\"display\":\"Clinical Not
e\"}]}],\"subject\":{\"reference\":\"Patient/patientid12345\"},\"date\":\"2022-04-13T05:37:38.982+00:00\",\"autho
r\":[{\"reference\":\"Practitioner?identifier=http://hl7.org/fhir/sid/us-npi|9999990259\",\"display\":\"Dr. Tandra3
34 Carter549\"}],\"custodian\":{\"reference\":\"Organization?identifier=https://ocktank.com/synthetic/synthea|f7ae4
97d-8dc6-3721-9402-43b621a4e7d2\",\"display\":\"PCP14023\"},\"content\":[{\"attachment\":{\"contentType\":\"text/pl
ain; charset=utf-8\",\"data\":\"IFRoZSBwYXRpZW50IGlzIGFuIDg2LXllYXItb2xkIGZlbWFsZSBhZG1pdHRlZCBmb3IgZXZhHVhdGlvbiB
vZiBhYmRvbWluYWwgcGFpbiBhbmQgYmxvb25IHN0b29scy4gVGhlIHBhdGllbnQgaGFzIGNvbG9aXMgYW5kIGFsc28gZGl2ZXJ0aWN1bGl0aXMsIH
VuZGVyZ29pbmcgdHJlYXRtZW50LiBEdXJpbmcgdGhlIGhvc3BpdGFsaXphdGlvbiwgdGhlIHBhdGllbnQgY29tcGxhaW5zIG9mIHNob3J0bmVzcyBvZ
iBicmVhdGgsIHdoaWNoIGlzIHdvcnNlbmluZy4gVGhlIHBhdGllbnQgdW5kZXJ3ZW50IGFuIGVjaG9jYXJkaW9ncmFtLCB3aGljaCBzaG93cyBzZXZl
cmUgbWl0cmFsIHJlZ3VyZ2l0YXRpb24gYW5kIGFsc28gbGFyZ2UgcGxldXJhbCBlZmZ1c2lvbi4gVGhpcyBjb25zdWx0YXRpb24gaXMgZm9yIGZ1cnR
oZXIgZXZhHVhdGlvbiBpbiB0aGlzIHJlZ2FyZC4gQXMgcGVyIHRoZSBwYXRpZW50LCBzaGUgaXMgYW4gODYteWVhci1vbGQgZmVtYWxlLCBoYXMgbGlt
aXRlZCBhY3Rpdml0eSBsZXZlbC4gU2hlIGhhcyBiZWVuIGhhdmluZyBzaG9ydG5lc3Mgb2YgYnJlYXRoIGZvciBtYW55IHllYXJzLiBTaGUgYWxzb
yB3YXMgdG9sZCB0aGF0IHNoZSBoYXMgYSBoZWFydCBtdXJtdXIsIHdoaWNoIHdhcyBub3QgZm9sbG93ZWQgdGhyb3VnaCBvbiBhIHJlZ3VsYXIgYmFz
aXMu\"},\"format\":{\"system\":\"http://ihe.net/fhir/ValueSet/IHE.FormatCode.codesystem\",\"code\":\"urn:ihe:iti:xd
s:2017:mimeTypeSufficient\",\"display\":\"mimeType Sufficient\"}}],\"context\":{\"encounter\":[{\"reference\":\"Enc
ounter/98309c16-75dd-8780-4bab-2d2788f7f885\"}],\"period\":{\"start\":\"2022-04-13T05:07:38+00:00\",\"end\":\"2022-
04-13T05:22:38+00:00\"}}}"

Figure 7.20 – DocumentReference JSON data

12. Next, we will ingest the DocumentReference FHIR resource into Amazon HealthLake. To do that, we create a resource endpoint, which includes our FHIR data store with the DocumentReference resource ID:

```
# Parse the input arguments
request_body_file = "doc_ref.json"
region = "us-east-1"
# Frame the resource endpoint
resource_endpoint = data_store_endpoint+resource_path
resource_endpoint = "<YOUR_DATASTORE_ENDPOINT>/r4/
DocumentReference/" + "id12345"
```

13. The FHIR API expects to have SigV4. We are using the `AWSSigV4` library and creating Sigv4 authentication:

```
!pip install requests-auth-aws-sigv4
from requests_auth_aws_sigv4 import AWSSigV4
# Frame authorization
auth = AWSSigV4("healthlake", session=session)
```

14. We now read the `doc_ref.json` file and store the data in the `json_data` object. Now we are calling the PUT FHIR API to store the DocumentReference FHIR resource in the Amazon HealthLake data store. We are passing `json_data` along with authentication token to the put method, as in `requests.put(resource_endpoint, json=json_data, auth=auth)`:

```
# Read the request body from input file
with open(request_body_file) as json_body:
    json_data = json.load(json_body)
# Calling data store FHIR endpoint using SigV4 auth
r = requests.put(resource_endpoint, json=json_data,
auth=auth)
print(r.json())
```

15. Let's print the response of the PUT request, as follows:

{'resourceType': 'DocumentReference', 'id': 'id12345', 'meta': {'lastUpdated': '2022-06-09T19:54:40.666Z', 'profile
': ['http://hl7.org/fhir/us/core/StructureDefinition/us-core-documentreference']}, 'identifier': [{'system': 'urn:i
etf:rfc:3986', 'value': 'urn:uuid:f11b9b48-ff6b-a62b-8425-037ea9c2d826'}], 'status': 'current', 'type': {'coding':
[{'system': 'http://loinc.org', 'code': '34117-2', 'display': 'History and physical note'}, {'system': 'http://loin
c.org', 'code': '51847-2', 'display': 'Evaluation+Plan note'}]}, 'category': [{'coding': [{'system': 'http://hl7.or
g/fhir/us/core/CodeSystem/us-core-documentreference-category', 'code': 'clinical-note', 'display': 'Clinical Note
'}]}], 'subject': {'reference': 'Patient/3cedeb4b-a9b8-41d2-aef2-de4c0413ec95'}, 'date': '2022-04-13T05:07:38.982+0
0:00', 'author': [{'reference': 'Practitioner?identifier=http://hl7.org/fhir/sid/us-npi|9999990259', 'display': 'D
r. Tandra334 Carter549'}], 'custodian': {'reference': 'Organization?identifier=https://ocktank.com/synthetic/synthe
a|f7ae497d-8dc6-3721-9402-43b621a4e7d2', 'display': 'PCP14023'}, 'content': [{'attachment': {'contentType': 'text/p
lain; charset=utf-8', 'data': 'IFRoZSBwYXRpZW50IGlzIGFuIDg2LXllYXItb2xkIGZlbWFsZSBhZG1pdHRlZCBmb3IgZXhhbHHVdGlvbiBv
ZiBhYmRvbWluYWwgcGFpbBhbmQgYmxvb2R5IHN0b29scy4gVGhlIHBhdGllbnQgaGFzIGNvbG0aXMgYW5kIGFsc28gZGl2ZXJ0aWN1bGl0aXMsIHV
uZGVyZ29pbmcgdHJlYXRtZW50LiBEdXJpbmcgdGhlIGhvc3BpdGFsaXphdGlvbiwgdGhlIHBhdGllbnQgY29tcGxhaW5zIG9mIHNob3J0bmVzcyBvZi
BicmVhdGgsIHdoaWNoIGlzIHdvcnNlbmluZy4gVGhlIHBhdGllbnQgdW5kZXJ3ZW50IGFuIGVjaG9jYXJkaW9ncmFtLCB3aGljaCBzaG93cyBzZXZlc
mUgbWl0cmFsIHJlZ3VyZ210YXRpb24gYW5kIGFsc28gbGVmdCBzZW50cmljdWxhciBheXBlcnRyb3BoeS4gVGhlIHBhdGllbnQgdW5kZXJ3ZW50IGFuIG
ZXIgZXZhbHVhdGlvbiBpbiB0aGlzIHJlZ2FyZC4gQXMgcGVyIHRoZSBwYXRpZW50LCBzaGUgaGFzIGW5YW5naW5hIGZsc28gZGl2ZXJ0aWN1bGl0aXM
taXRlZCBhY3Rpdml0eSBSZXZlbC4gU2h1IGhhcyBiZWVuIGhhdmluZyBzaG9ydG5lc3Mgb2YgYnJlYXRoIGZvciBYW55IHllYXJzLiBTaGUgYWxzby
B3YXMgdG9sZCB0aGF0IHNoZSBoYXMgYSBoZWFydCBtdXJtdXIsIHdoaWNoIHdhcyBub3QhcyBub3QgZm9sbG93ZWQgdGhyb3VnaCBiBIHJlZ3VsYXIgYmFza
XMu'}, 'format': {'system': 'http://ihe.net/fhir/ValueSet/IHE.FormatCode.codesystem', 'code': 'urn:ihe:iti:xds:201
7:mimeTypeSufficient', 'display': 'mimeType Sufficient'}}], 'context': {'encounter': [{'reference': 'Encounter/9830
9c16-75dd-8780-4bab-2d2788f7f885'}], 'period': {'start': '2022-04-13T05:07:38+00:00', 'end': '2022-04-13T05:22:38+0
0:00'}}}}

Figure 7.21 – Document reference JSON data – JSON file

We learned how to convert an unstructured document to an FHIR resource and then upload it to Amazon HealthLake to create a centralized view of patient health data.

Now, let's summarize the chapter.

Summary

In this chapter, we discussed the fundamentals of FHIR and how to use it in the healthcare industry to solve challenges such as healthcare data interoperability. We also discussed Amazon HealthLake and its core features for storing, transforming, and analyzing health data. Amazon HealthLake's NLP models interpret medical insights such as medical condition, medication, dosage, medical ontology linking, and more from health data, which can be further leveraged to create additional models with Amazon SageMaker or visualizations.

We then walked through the console and code to see how to create an Amazon HealthLake FHIR data store and how to input FHIR resources into our data store. We also discussed a sample architecture and implementation to ingest document-based health data into Amazon HealthLake to create a centralized, secure, scalable, HIPAA-eligible health data lake.

In the next chapter, we will extend the discussion to healthcare data interoperability. We will see how we can use AWS AI services with scalable AWS infrastructure to define a healthcare data interoperability platform.

References

- *To find out more about HL7, visit*: https://www.hl7.org/fhir/.
- *To find out more about FHIR operations, visit*: https://www.hl7.org/fhir/operations.html.

Part 3: Intelligent Document Processing in Industry Use Cases

In this part, we will focus on how to build an IDP pipeline for the healthcare industry. Moreover, we will learn how to create a secure, HIPAA-eligible IDP solution. Then, we will learn how to build an IDP pipeline for insurance processing and mortgage application processing.

This section comprises the following chapters:

8

IDP Healthcare Industry Use Cases

In the previous chapter, you understood what **Fast Healthcare Interoperability Resources (FHIR)** is and how Amazon HealthLake can help with FHIR. Moreover, we looked into how we can manage unstructured non-FHIR documents to store in a centralized data store with an **Intelligent Document Processing** (**IDP**) pipeline. We will now go into more detail about FHIR and see how it helps in healthcare interoperability, with various industry use cases. We will also check how Amazon IDP helps in healthcare industry use cases. You will learn how to build an IDP pipeline for the healthcare industry. Moreover, you will learn how to create a secure, compliant IDP solution.

We will navigate through the following sections in this chapter:

- Understanding IDP with healthcare prior authorization
- Learning IDP for pharmacy receipt automation
- Understanding healthcare claims processing and risk adjustment with IDP

Technical requirements

For this chapter, you will need access to an AWS account. Before getting started, we recommend that you create an AWS account by referring to the AWS account setup and Jupyter notebook creation steps in the *Technical requirements* section in *Chapter 2, Document Capture and Categorization*.

You can find Chapter-8 code sample in GitHub: `https://github.com/PacktPublishing/Intelligent-Document-Processing-with-AWS-AI-ML-/tree/main/chapter-8`. Also recommend to check availability of AI service in your AWS regions before using it.

Understanding IDP with healthcare prior authorization

In this section, we will see how to use AWS AI services to accelerate the healthcare prior authorization process, but before diving deep into AWS AI services, let's learn about the healthc are prior authorization process.

An introduction to the healthcare prior authorization process

What is "prior authorization?" Prior authorization is a process to receive approval before any healthcare treatments, services, drugs, and so on. Most often, patients or providers request approval from a payer prior to any type of healthcare service so that the cost of it will be covered by the insurance plan. Healthcare insurance companies use the process of prior authorization to check for the right financial responsibility to keep healthcare cost-effective for patients. However, during the prior authorization process, there are many manual transactions of documents, making the process slow, time-consuming, and inefficient, resulting in a delay in receiving treatments, leading to patients not having the right value-based care. For example, during the prior authorization process, the provider manually enters data on a payer-specific portal, or exchanges documents by fax. Most often, manual data entry and the verification process require human effort, which becomes time-consuming and expensive. Can we help automate this prior authorization process with AWS IDP? To answer this, let's look at the AWS infrastructure for data exchange and AWS AI services to find a solution for the prior authorization use case.

In the previous chapter, we looked into how the future of a healthcare data exchange depends on a common language, which can enable communication across different healthcare systems, such as a payer or provider. This seamless data exchange at a high level can be achieved with healthcare interoperability. The not-for-profit organization **Health Level Seven** (**HL7**) **International** has been a leader in defining healthcare data interoperability since 1987. HL7 has developed healthcare data modeling for **Electronic Health Records** (**EHRs**) to enable a meaningful exchange of healthcare information, with data kept confidential and secure. Its latest framework, FHIR, is a step forward to make the process seamless, with two broader categories. This includes a data model for healthcare information and a technology stack for application development. This was developed in collaboration with an open source community for wide adoption and access. Since 2012, FHIR has accelerated the development of healthcare data interoperability, allowing faster, easier healthcare data sharing. Health data is growing at a rapid pace. At the same time, technology is advancing at a rapid pace. This growing health data, with advanced technology, has created the requirement for consistent healthcare data interoperability across healthcare systems.

We have discussed how healthcare data interoperability (in *Chapter 7, Accurate Extraction and Health Insights with Amazon HealthLake*) can help with FHIR data exchange.

In 2018, the US Office of the National Coordinator for Health Information Technology launched a testing program called *Inferno* to support the adoption of FHIR. Inferno allows developers to verify that their FHIR standards are consistently implemented across systems. This will help ensure that developers adhere to FHIR standards and that communication remains consistent. Moreover, **Centers for Medicare and Medicaid Services** (**CMS**) released a rule on coverage transparency, helping the wider adoption of a healthcare data exchange.

Here is a link to dive into details about the health plan price transparency: `https://www.cms.gov/healthplan-price-transparency`, and here is a link to dive into details about FHIR: `https://www.hl7.org/fhir/`. Paired with cloud technologies, FHIR is turning real-time interoperability into a reality. Data centers that held key medical data were difficult to scale, maintain, and access. Now, with FHIR operating as a common language of a data exchange, we can implement a healthcare interoperability platform.

Community collaboration is also shaping a data exchange platform for healthcare. One such community, Da Vinci, has developed reference implementations for FHIR healthcare data interoperability (`https://github.com/HL7-DaVinci`). You can extend these reference implementations to find a solution for healthcare prior authorization use cases, with **Coverage Requirements Discovery** (**CRD**) as an implementation example. Automated prior authorization can help speed up processes, as well as reduce the burden on the provider of additional administrative work, resulting in increased value-based care for the patient. Let's look at a quick architectural guidance for CRD for prior authorization use cases, as shown here.

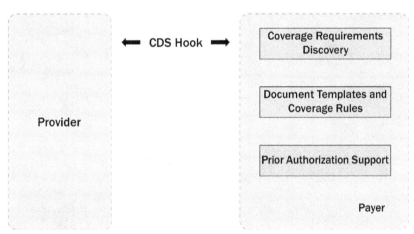

Figure 8.1 – Healthcare prior authorization – block architecture

The preceding block architecture shows the communication of two components – in this case, a payer and provider healthcare data exchange with CDS hooks. It also describes the components required by the payer to implement the framework. In *Figure 8.2*, you can also check out a reference architecture to deploy this prior authorization interoperability platform on an AWS Cloud container platform.

The detailed implementation is out of the scope of this book. For additional information, we recommend that you check out the details here: `https://aws.amazon.com/blogs/industries/healthcare-simplify-authorization-aws-hl7-fhir/`.

You can find more details on CDS hooks here: `https://cds-hooks.hl7.org/`.

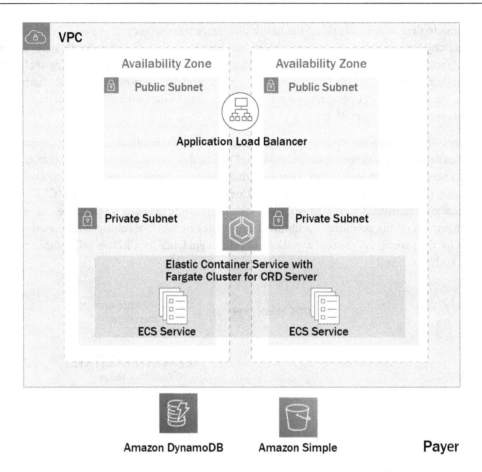

Figure 8.2 – An AWS prior authorization infrastructure – reference architecture

We just looked at reference architecture for prior authorization. Now, let's see another example of automating the prior authorization form-filling process with the help of IDP.

Automate prior authorization form filling using Amazon HealthLake

We have listed the pain point of the healthcare prior authorization process in the *An introduction to the healthcare prior authorization process* section. Can we help accelerate this prior authorization process with innovation? The answer is yes. One such technical implementation will be illustrated in the example of automating the prior authorization form-filling process. To automate the form-filling process, we need to extract the required information from clinical data. Let's now check out a sample prior authorization form, as shown here:

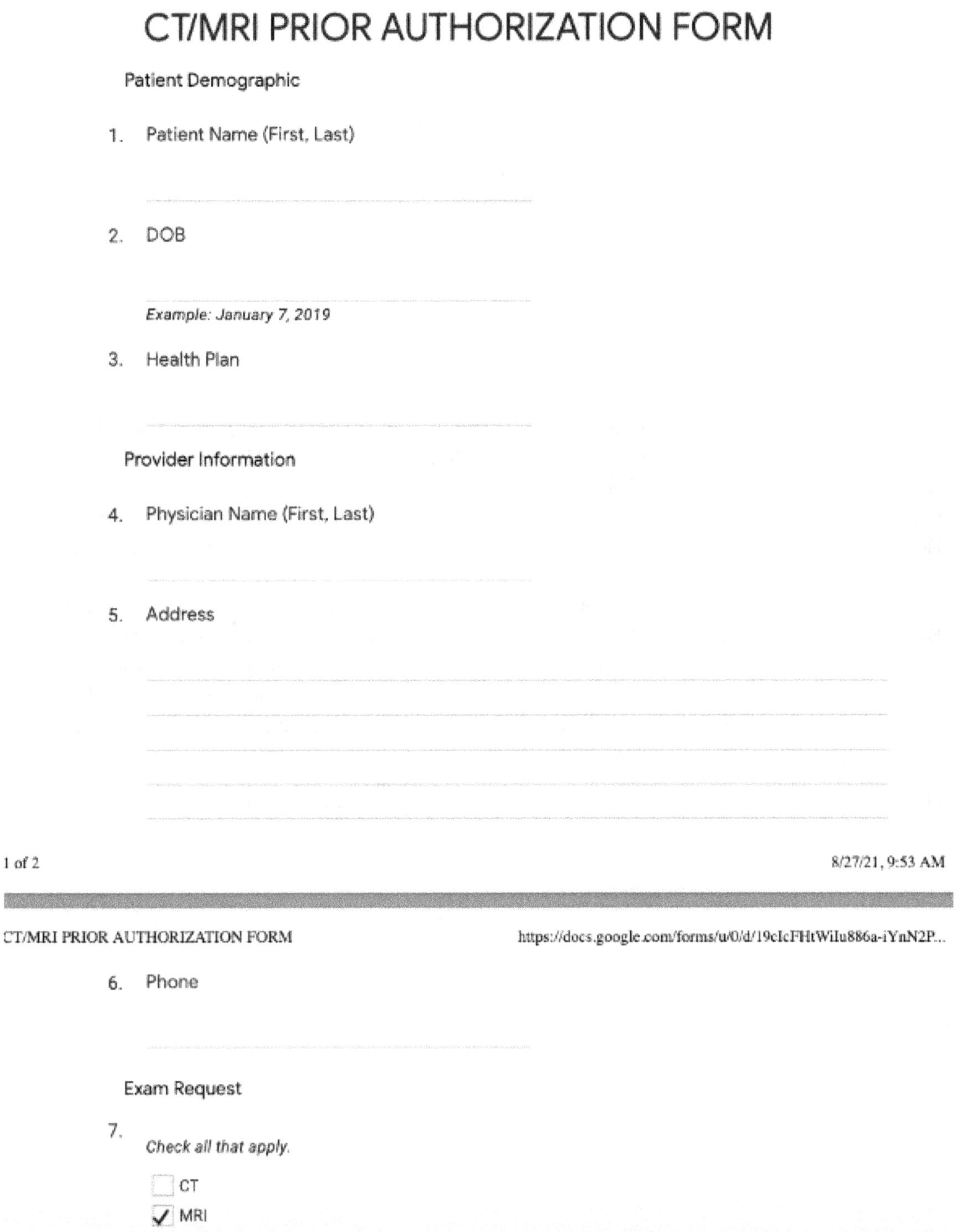

CT/MRI PRIOR AUTHORIZATION FORM

Patient Demographic

1. Patient Name (First, Last)

2. DOB

   Example: January 7, 2019

3. Health Plan

Provider Information

4. Physician Name (First, Last)

5. Address

1 of 2 8/27/21, 9:53 AM

CT/MRI PRIOR AUTHORIZATION FORM https://docs.google.com/forms/u/0/d/19cIcFHtWiIu886a-iYnN2P...

6. Phone

Exam Request

7. *Check all that apply.*

   - [ ] CT
   - [x] MRI

Figure 8.3 – A sample prior authorization form

To fill in this prior authorization form, we will use the following sample reference architecture:

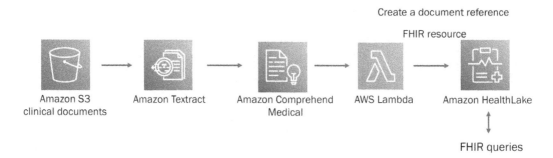

Figure 8.4 – Prior authorization architecture

For this reference architecture, we are assuming that we have an FHIR resource such as **Claim**, **Patient**, **Organization**, or **Condition** in our Amazon HealthLake data store. Also, we have additional clinical information in a non-FHIR format, such as a provider note in a PDF format. We convert a RAW document to a `DocumentReference` FHIR resource, and store it in the same Amazon HealthLake data store. If you do not know how to generate a `DocumentReference` FHIR resource from a non-FHIR clinical document type, refer to *Chapter 6, Review and Verification of Intelligent Document Processing*. Once we have all the clinical information regarding a patient in an Amazon HealthLake data store, we can use Amazon analytics services such as Amazon Athena and AWS Glue for quick analysis. We can query to derive required information such as patient name, date-of-birth information from the Patient FHIR resource, the health plan from Claim, and health plan address from Organization FHIR resources. We can also validate the accuracy of these elements with the Document Reference FHIR resource.

Let's look at a sample filled-in prior authorization form with these extracted elements:

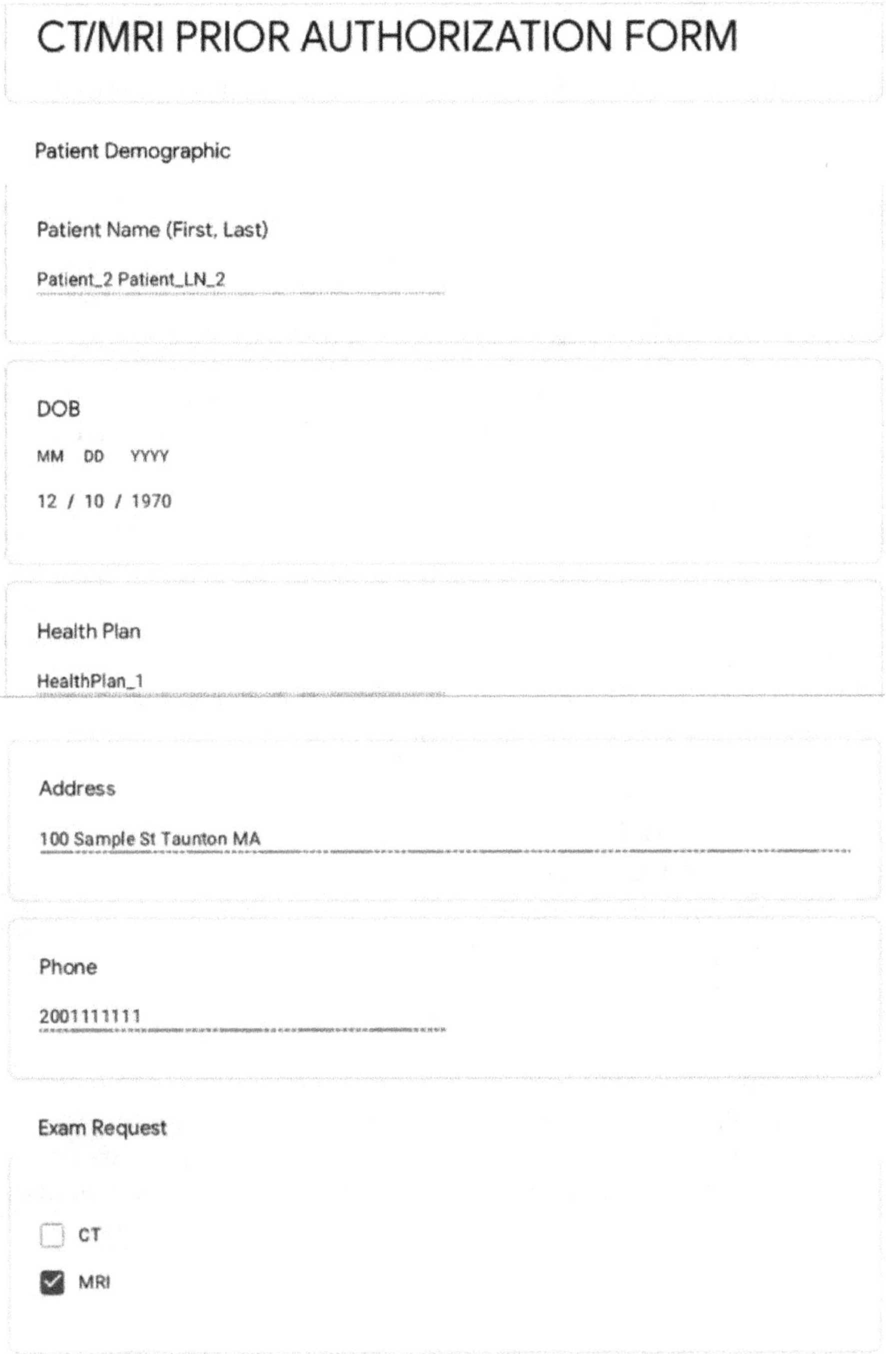

CT/MRI PRIOR AUTHORIZATION FORM

Patient Demographic

Patient Name (First, Last)

Patient_2 Patient_LN_2

DOB

MM   DD   YYYY

12 / 10 / 1970

Health Plan

HealthPlan_1

Address

100 Sample St Taunton MA

Phone

2001111111

Exam Request

☐ CT

☑ MRI

Figure 8.5 – A sample filled prior authorization form

We went through the reference architecture of how we can leverage AWS AI services to automate the prior authorization form-filling process. Now, let's look at IDP in another healthcare use case, pharmacy receipt processing. Note that I would recommend checking out the following links about designing a HIPAA-compliant architecture in AWS:

```
https://d1.awsstatic.com/Industries/HCLS/Resources/Architecting%20
for%20HIPAA%20one-pager%202018.pdf
```

```
https://aws.amazon.com/compliance/hipaa-compliance/
```

Learning IDP for pharmacy receipt automation

Now, we will dive into a solution to extract the required information from a prescription, levering our AWS IDP pipeline. For this exercise, we will use the following sample pharmacy prescription:

MANITOBA Pharmacy

204 Manitoba Street
Winnipeg MB M2B 2Y2 Canada
Store # 0001   Phone: 204-204-2004

Rx#2042042   Ref:0 Dr.Manitoba
Toba Man
    TAKE 1 CAPSULE THREE
    TIMES DAILY UNTIL
    FINISHED (ANTIBIOTIC)|
APO-AMOXI 500MG
AMOXICILLIN 500MG
RED/YEL/ELLIP/APO {500}
30 CAP   14 Oct 2007   Total 21.43

Figure 8.6 – A sample prescription

Our goal is to find out the important information to be filled in on a medical prescription. We will answer some key questions to determine what goes into the medical prescription. We will leverage the **Amazon Textract Queries** feature to find the answer to our questions. Amazon Textract Queries leverages a combination of visual, spatial, and language contexts to provide answers to our natural language questions.

Let's check out Amazon Textract Queries on the AWS Management Console for this sample prescription:

1. Go to Amazon Textract on the AWS Management Console. Click on **Try Amazon Textract**:

Figure 8.7 – The AWS Management Console – Try Amazon Textract

2. Click on **Choose document** and select the aforementioned sample prescription document:

Figure 8.8 – The Amazon Textract upload document screen

3. Select **Forms** and **Queries**, and add queries to the **Queries** section, as shown in the following screenshot:

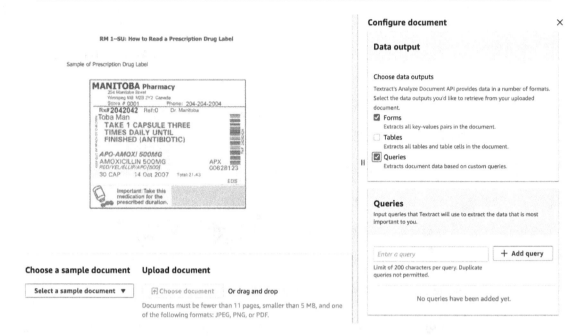

Figure 8.9 – Amazon Textract – Queries

Once you click **Apply configuration**, you can check the answers to our queries, as shown here:

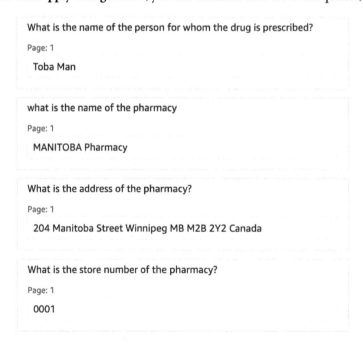

Figure 8.10 – Textract Queries on the AWS Management Console

In the following screenshot, we ask questions, leveraging Textract's Queries feature, about the physician's name, the prescription number, and the main ingredient of the prescription, and gain responses:

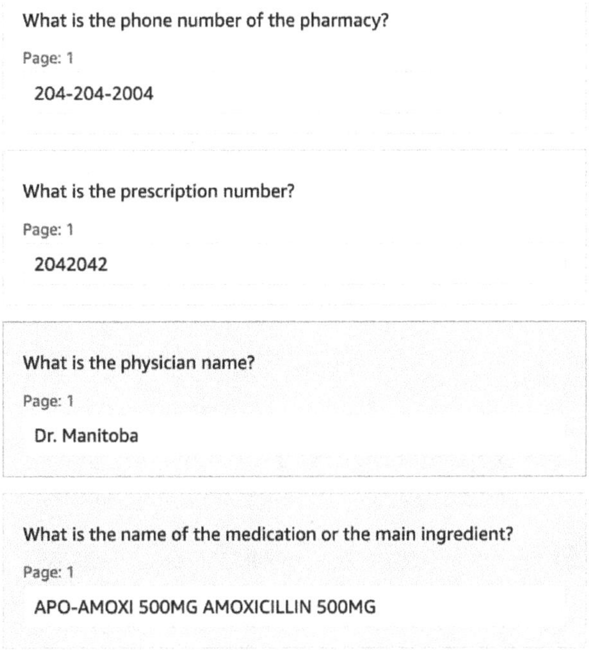

What is the phone number of the pharmacy?

Page: 1

204-204-2004

What is the prescription number?

Page: 1

2042042

What is the physician name?

Page: 1

Dr. Manitoba

What is the name of the medication or the main ingredient?

Page: 1

APO-AMOXI 500MG AMOXICILLIN 500MG

Figure 8.11– Additional Textract Queries responses on the AWS Management Console

In the following screenshot, we also ask questions, leveraging Textract Queries, about the date the prescription was filled, the dosage of the medication, and the directions to take it, and gain responses:

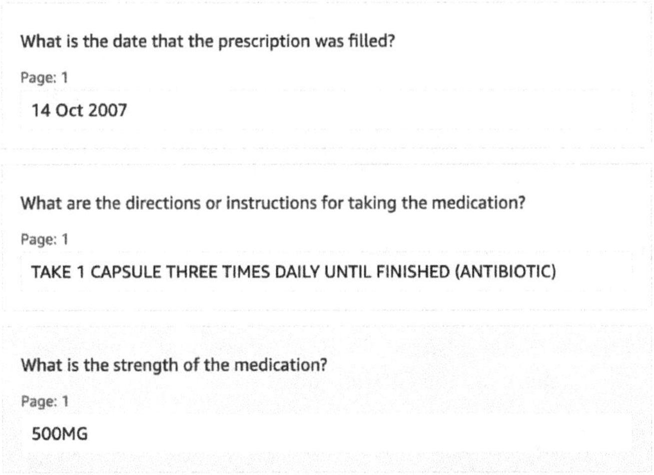

What is the date that the prescription was filled?

Page: 1

14 Oct 2007

What are the directions or instructions for taking the medication?

Page: 1

TAKE 1 CAPSULE THREE TIMES DAILY UNTIL FINISHED (ANTIBIOTIC)

What is the strength of the medication?

Page: 1

500MG

Figure 8.12 – An Amazon Textract Queries answer

We looked at how to leverage Textract Queries to extract information from a prescription document for drug refill. We will now run through the following sample code to automate the drug fill information from the prescription document programmatically.

You can have the full code walkthrough with the Notebook: `https://github.com/PacktPublishing/Intelligent-Document-Processing-with-AWS-AI-ML-/blob/main/chapter-8/healthcare-08.ipynb`:

1. Import the required libraries to run the sample code for accurate extraction of the prescription:

```
import boto3
import json
import re
import io
from io import BytesIO
from pprint import pprint
from IPython.display import Image, display
from PIL import Image as PImage, ImageDraw
```

2. Install the Textract response parser for an easier JSON response parse:

```
!pip install amazon-textract-response-parser
```

3. We will use the prescription image from *Figure 8.6*. You can check the sample image with the following code:

```
from IPython.display import Image
image_filename = "prescription.png"
Image(filename=image_filename)
```

4. Get the `boto3` Textract client:

```
#create a Textract Client
textract = boto3.client('textract')
# Document
documentName = image_filename
```

5. We will call the `analyze_document` API with the Textract `FeatureTypes` value as `Queries` and `Forms`. We will also pass in the required questions in `QueriesConfig`. For a full list of questions, check out the GitHub code:

```
response = None
with open(documentName, 'rb') as document:
```

```
        imageBytes = bytearray(document.read())
        # Call Textract
        response = textract.analyze_document(
            Document={'Bytes': imageBytes},
            # new QUERIES Feature Type for Textract Queries
            # We could add additional Feature Types like
FORMS and/or TABLES
            # FeatureTypes=["QUERIES", "FORMS", "TABLS"],
            FeatureTypes=["QUERIES"],
            QueriesConfig={
                "Queries": [{
                    "Text": "What is the name of the
pharmacy?",
                    "Alias": "PHARMACY_NAME"
                },
                {
                    "Text": "What is the address of the
pharmacy?",
                    "Alias": "PHARMACY_ADDRESS"
                },
                {
                    "Text": "What is the store number of the
pharmacy",
                    "Alias": "PHARMACY_STORE_NUM"
                }
            ]
        })
        print(response)
```

6. Now, we can see some sample extracted answers for our Textract Queries results:

9.7336196899414, 'Text': 'duration.', 'TextType': 'PRINTED', 'Geometry': {'BoundingBox': {'Width': 0.14376130700111
39, 'Height': 0.0345890037715435, 'Left': 0.3581547439098358, 'Top': 0.9218687415122986}, 'Polygon': [{'X': 0.35815
47439098358, 'Y': 0.9218687415122986}, {'X': 0.5019160509109497, 'Y': 0.9218687415122986}, {'X': 0.501916050910949
7, 'Y': 0.9564577341079712}, {'X': 0.3581547439098358, 'Y': 0.9564577341079712}]}, 'Id': '959f4d88-909f-4cf0-954f-2
3d8be6ab543'}, {'BlockType': 'QUERY', 'Id': 'deefc40a-6a78-4e05-b473-ae1d1d6f16a1', 'Relationships': [{'Type': 'ANS
WER', 'Ids': ['d3867c56-c78a-4768-b3fd-48211cb6b405']}], 'Query': {'Text': 'What is the name of the pharmacy?', 'Al
ias': 'PHARMACY_NAME'}}, {'BlockType': 'QUERY_RESULT', 'Confidence': 90.0, 'Text': 'MANITOBA Pharmacy', 'Geometry':
{'BoundingBox': {'Width': 0.5208667516708374, 'Height': 0.06438631564378738, 'Left': 0.024077046662569046, 'Top':
0.03018108569085598}, 'Polygon': [{'X': 0.024077046662569046, 'Y': 0.03018108569085598}, {'X': 0.5449438095092773,
'Y': 0.03018108569085598}, {'X': 0.5449438095092773, 'Y': 0.09456740319728851}, {'X': 0.024077046662569046, 'Y': 0.
09456740319728851}]}, 'Id': 'd3867c56-c78a-4768-b3fd-48211cb6b405'}, {'BlockType': 'QUERY', 'Id': '9058cf45-467e-4c

Figure 8.13 – An Amazon Textract Queries response

We walked through some sample code to extract information from the prescription for automatic drug fill.

Understanding healthcare claims processing and risk adjustment with IDP

When a medical claim is submitted, the insurance provider (payer) must process the claim to determine their correct financial responsibility and that of the patient. The process is known as claims adjudication or claims processing. During this process, the claims go through various checks; one such check is coding-level validation. During this coding-level validation, we check for the accuracy of the medical diagnosis code. Most often, this step requires manual review by a medical professional. Can we automate this process by leveraging technology? The answer is yes.

We will use Amazon IDP for coding-level validation for accurate claims processing. For this exercise, we will use the CMS1500 claim form and also a doctor's note. We will process these documents with Amazon Textract for the accurate extraction of elements along with the ICD-10-CM code. We will also process the doctor's note to adhere to the ICD-10-CM code, and finally, we can check whether the code mentioned in the healthcare application form matches the code inferred from the doctor's note. Now, let's dive into the detailed implementation:

1. We will use cms100-1.png as a sample document for this exercise. You can run the following code to check the sample image:

    ```
    # Document
    documentName = "cms1500-1.png"
    display(Image(filename=documentName))
    ```

2. Check the sample result of the CMS application form:

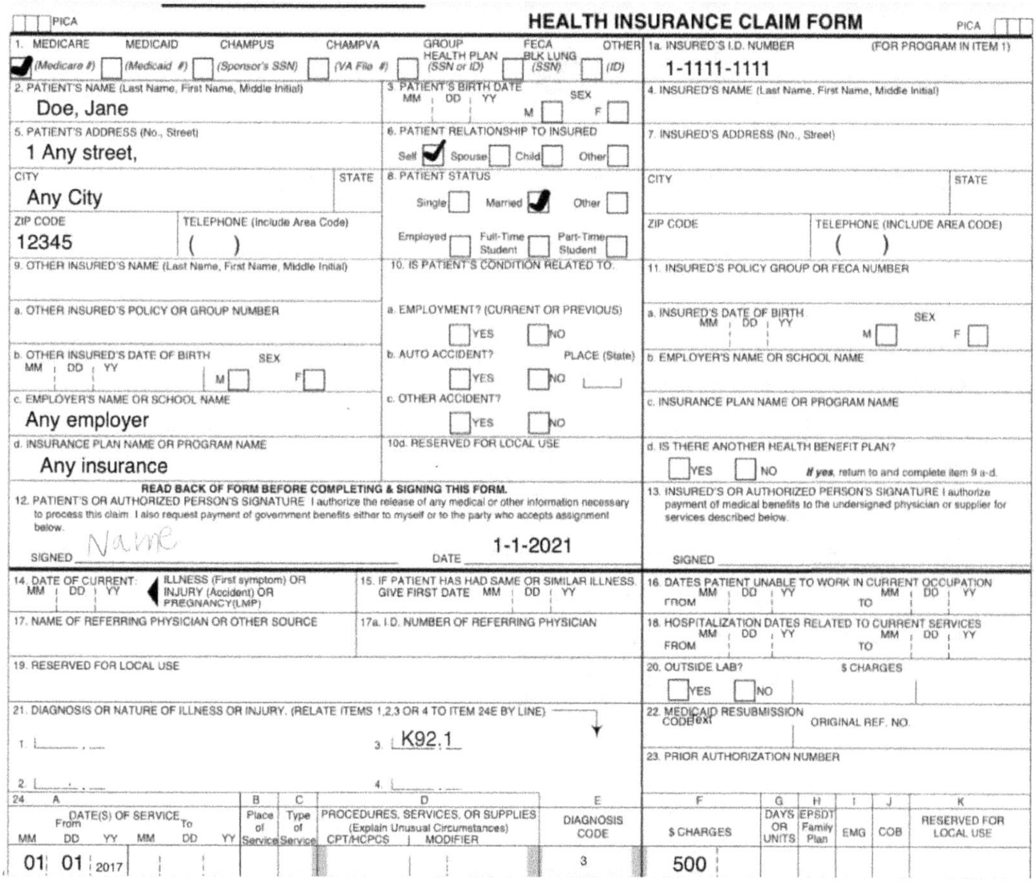

Figure 8.14 – The sample CMS1500 form

3. We have defined a method name, `calltextract`, to call the Amazon Textract `analyze_document` method with RAW image bytes:

```
# process using image bytes
def calltextract(documentName):
    client = boto3.client(service_name='textract',
        region_name= 'us-east-1',
        endpoint_url='https://textract.us-east-1.
amazonaws.com')
    with open(documentName, 'rb') as file:
        img_test = file.read()
        bytes_test = bytearray(img_test)
        print('Image loaded', documentName)
```

```
      # process using image bytes
      response = client.analyze_document(Document={'Bytes':
bytes_test}, FeatureTypes=['FORMS'])
      return response
```

4. Now, call the `calltextract` method defined in the preceding code block:

```
response= calltextract(documentName)
print(response)
```

5. We have defined a method name, `getformkeyvalue()`, to extract key value information from the JSON response of Amazon Textract:

```
#Extract key values
# Iterate over elements in the document
from trp import Document
def getformkeyvalue(response):
    doc = Document(response)
    key_map = {}
    for page in doc.pages:
        # Print fields
        for field in page.form.fields:
            if field is None or field.key is None or
field.value is None:
                continue
            key_map[field.key.text] = field.value.text
    return key_map
```

6. Now, call the `getformkeyvalue` method to parse the Textract JSON response to get all key-value pairs:

```
get_form_keys = getformkeyvalue(response)
print(get_form_keys)
```

7. You can find the results of all the key-value pairs from the sample document of the CMS application form. Note that the ICD-10-CM code extracted from the application is K92.1:

```
{'YES': 'NOT_SELECTED', 'NO': 'NOT_SELECTED', 'ZIP CODE': '12345', 'Full-Time
Student': 'NOT_SELECTED', 'CITY': 'Any City', "5. PATIENT'S ADDRESS (No.,
Street)": '1 Any street,', 'Part-Time Student': 'NOT_SELECTED', "c.
EMPLOYER'S NAME OR SCHOOL NAME": 'Any employer', "2. PATIENT'S NAME (Last
Name, First Name, Middle Initial)": 'Doe, Jane', 'Other': 'NOT_SELECTED',
'Single': 'NOT_SELECTED', 'TELEPHONE (Include Area Code)': '( )', 'M':
'NOT_SELECTED', 'TELEPHONE (INCLUDE AREA CODE)': '( )', 'Self': 'SELECTED',
'Child': 'NOT_SELECTED', 'd. INSURANCE PLAN NAME OR PROGRAM NAME': 'Any
insurance', 'Spouse': 'NOT_SELECTED', "(Sponsor's SSN)": 'NOT_SELECTED',
'DATE': '1-1-2021', '(Medicaid #)': 'NOT_SELECTED', 'F': 'NOT_SELECTED',
'Married': 'SELECTED', '(Medicare #)': 'SELECTED', 'PICA': 'NOT_SELECTED',
'(ID)': 'NOT_SELECTED', '3.': 'K92.1', 'SIGNED': 'Name', 'FECA BLK LUNG
(SSN)': 'NOT_SELECTED', "1a. INSURED'S I.D. NUMBER": '1-1111-1111', '(VA File
#)': 'NOT_SELECTED'}
```

Figure 8.15 – A sample Textract form response

8. Now, we will process another document type, which is the doctor's note. You can use the following code to check the image:

```
documentName = "doctornotes1.png"
display(Image(filename=documentName))
```

The following is a sample image of the RAW doctor's note:

The patient is an 86-year-old female admitted for evaluation of abdominal pain and bloody stools. The patient has colitis and also diverticulitis, undergoing treatment. During the hospitalization, the patient complains of shortness of breath, which is worsening. The patient underwent an echocardiogram, which shows severe mitral regurgitation and also large pleural effusion. This consultation is for further evaluation in this regard. As per the patient, she is an 86-year-old female, has limited activity level. She has been having shortness of breath for many years. She also was told that she has a heart murmur, which was not followed through on a regular basis.

Figure 8.16 – A sample provider note

9. We are calling into `calltextract()`, as defined in *step 3* with the doctor's note document type. We collected all the lines from the doctor's note for further processing by Amazon Comprehend Medical:

```
response= calltextract(documentName)
# Print text
print("\nText\n========")
text = ""
for item in response["Blocks"]:
    if item["BlockType"] == "LINE":
        print ('\033[94m' +  item["Text"] + '\033[0m')
        text = text + " " + item["Text"]
```

10. We are creating a Boto3 `comprehendmedical` client and calling into the `Comprehend Medical detect_entities` API. We are passing in the Textract response in *step 10* as input to this API. Amazon Comprehend Medical extracts all the medical entities from this document:

```
comprehend = boto3.client(service_
name='comprehendmedical')
# Detect medical entities
cm_json_data =  comprehend.detect_entities_v2(Text=text)
print("\nMedical Entities\n========")
for entity in cm_json_data["Entities"]:
    print("- {}".format(entity["Text"]))
    print ("   Type: {}".format(entity["Type"]))
    print ("   Category: {}".format(entity["Category"]))
    if(entity["Traits"]):
        print("   Traits:")
        for trait in entity["Traits"]:
            print ("    - {}".format(trait["Name"]))
    print("\n")
```

11. You can check out the Amazon Comprehend Medical-extracted entities from this provider note, as shown here:

```
Medical Entities
========
- 86
    Type: AGE
    Category: PROTECTED_HEALTH_INFORMATION


- abdominal
    Type: SYSTEM_ORGAN_SITE
    Category: ANATOMY


- abdominal pain
    Type: DX_NAME
    Category: MEDICAL_CONDITION


- bloody stools
    Type: DX_NAME
    Category: MEDICAL_CONDITION
    Traits:
      - SIGN


- colitis
    Type: DX_NAME
    Category: MEDICAL_CONDITION
    Traits:
      - DIAGNOSIS


- diverticulitis
    Type: DX_NAME
    Category: MEDICAL_CONDITION
    Traits:
      - DIAGNOSIS


- treatment
    Type: TREATMENT_NAME
    Category: TEST_TREATMENT_PROCEDURE


- shortness of breath
    Type: DX_NAME
    Category: MEDICAL_CONDITION


- echocardiogram
    Type: TEST_NAME
    Category: TEST_TREATMENT_PROCEDURE


- mitral
    Type: SYSTEM_ORGAN_SITE
    Category: ANATOMY
```

Figure 8.17 – An Amazon Comprehend Medical response

12. We can leverage Amazon Comprehend Medical's inferred `_icd10_cm` method to infer a possible ICD-10-CM code from the provider note. Amazon Comprehend Medical also gives a confidence score. We are using a confidence score threshold of 90% to filter the response:

```
cm_json_data =  comprehend.infer_icd10_cm(Text=text)
print("\n Medical coding\n========")
for entity in cm_json_data["Entities"]:
    for icd in entity["ICD10CMConcepts"]:
        if (icd["Score"] >= 0.90):
            code = icd["Code"]
            print(code)
```

You can check out the extracted medical diagnosis code by Amazon Comprehend Medical here:

```
Medical coding
========
K92.1
R06.02
J90
R06.02
R01.1
```

Figure 8.18 – Amazon Comprehend Medical coding

You can see from Amazon Comprehend Medical that one of the inferred ICD-10-M codes is K92.1, which matches the ICD-10-CM code mentioned in our CMS1500 application form. This can be used as a quick mechanism to validate medical code in claims processing use cases.

The inferred ICD-10-CM code can be further mapped to a **Hierarchical Condition Category (HCC)** code by taking into account additional demographic information. This can be useful in healthcare risk adjustment use cases.

Summary

In this chapter, we discussed the IDP pipeline and how this can apply to healthcare industry use cases, such as healthcare prior authorization, prescription automation, and healthcare claims processing with health risk adjustment.

We then dove deep into healthcare prior authorization use cases with a coverage requirement request and its reference architecture on AWS. We also discussed how to automate filling in a prior authorization form from a clinical data store. Moreover, we discussed how we can automate drug fill information by automating extraction from a prescription document. Finally, we looked at how to process a document for healthcare claims processing with a risk adjustment use case.

In the next chapter, we will extend IDP to additional industry use cases, such as insurance. Moreover, we will dive deep into insurance claims processing use cases and see how IDP can help to automate claims processing.

9

Intelligent Document Processing – Insurance Industry

In the previous chapter, you understood the challenges involved in the healthcare industry to process documents. The healthcare industry is evolving to exchange meaningful data following healthcare interoperability standards, such as **Fast Healthcare Interoperability Resources** (**FHIR**), but most often, we need to handle unstructured documents in a non-FHIR format. We looked into a solution to bring these documents into an FHIR format for healthcare data interoperability. We also discussed a potential solution with AWS AI services such as Amazon Textract, Amazon Comprehend Medical, and Amazon HealthLake to automate this **Intelligent Document Processing** (**IDP**) pipeline. We dived into each stage of IDP in *Chapter 1, Intelligent Document Processing with AWS AI and ML* through *Chapter 7, Accurate Extraction and Health Insights with Amazon HealthLake*. We will now change gear and look into how we can apply these IDP stages to use cases in the insurance industry to process documents.

We will also dive into the IDP pipeline and how IDP can help automate use cases in the insurance industry. We will navigate through the following sections in this chapter:

- Automating the benefits enrollment process with IDP
- Understanding insurance claims processing extraction with IDP
- Understanding insurance claims processing document enrichment and review and verification

Technical requirements

For this chapter, you will need access to an AWS account. Before getting started, we recommend that you create an AWS account by referring to the AWS account setup and Jupyter notebook creation steps mentioned in the *Technical requirements* section in *Chapter 2, Document Capture and Categorization*.

You can find Chapter-9 code sample in GitHub: `https://github.com/PacktPublishing/Intelligent-Document-Processing-with-AWS-AI-ML-/tree/main/chapter-9`. Also recommend to check availability of AI service in your AWS regions before using it.

We have already learned about the stages of Intelligent Document Processing and how AWS AI services can help automate processing documents in each stage of IDP. Just to recap, we went through the following stages in *Chapter 1* through *Chapter 7*:

Figure 9.1 – IDP pipeline

Now, we will apply these IDP stages to specific use cases in the insurance industry:

- Automating the benefit enrollment application process
- Automating the insurance claims adjudication process

Automating the benefits enrollment process with IDP

Before diving deep, let's first understand how insurance companies work and what an insurance underwriting process is.

Insurance underwriting is a process that insurance companies follow to assess the risk and profit of offering a policy. For that reason, insurance companies define ways to decide how much of a risk it is to provide coverage. They need to know the possibility of something going wrong by checking the identity of a person and a person's history or records. To verify these details, insurance companies collect additional data and supporting documents. After the review and verification stage and calculating the risk, the underwriter sets the premium for the insurance.

For example, suppose someone is looking for car insurance. An insurance underwriter will evaluate their identity and driving history and check the details of the vehicle to determine the insurance policy details. During the process, they evaluate the risk score of the customer and determine the policy premium accordingly. Similarly, at the time of claims submission, the case worker evaluates the claims for their validity and authenticity to determine the correct financial responsibility.

Now we understand the underwriting process and why the insurance industry follows it before enrolling a benefit application. The underwriter is assessing the benefit the application is for. The underwriter is assessing the benefit the application is for. During this process, insurance industries had to process not just the benefit application document but also a large number of supporting documents. We will now dive into this benefit enrollment use case and use AWS AI services to automate the process. During the benefit enrollment process, the user submits the application to request a benefit; for example, a user submits an application for food stamps or to get a healthcare benefit. In this enrollment process, additional documents and data are required to check the benefit eligibility of the person. Most often, data is collected manually and reviewed across documents to check for the eligibility of the benefit. We will look into an automation solution to automate the benefit enrollment process with AWS IDP services such as Amazon Textract and Amazon Comprehend. Let's now dive into the following sample code to automate these steps. We will only focus on the data capture and data extraction stages of the IDP pipeline for the implementation:

For full code walk-through follow the steps in the Notebook: `https://github.com/ PacktPublishing/Intelligent-Document-Processing-with-AWS-AI-ML-/ blob/main/chapter-9/claims-processing-09.ipynb`

1. First, we import the required dependencies to run the code:

```
import boto3
import json
import boto3
import csv
import sagemaker
from sagemaker import get_execution_role
from sagemaker.s3 import S3Uploader, S3Downloader
from pprint import pprint
from IPython.display import Image, display
from PIL import Image as PImage, ImageDraw
```

2. We are using the Textract response parser, as follows, to parse the Textract response:

```
!pip install amazon-textract-response-parser
```

3. We are using a session bucket and IAM role for the execution:

```
role = get_execution_role()
#print("RoleArn: {}".format(role))
sess = sagemaker.Session()
bucket = sess.default_bucket()
prefix = 'claims-process-textract'
```

4. As we always start with a document. For this sample code, we are using a benefit application (form 297). You can use the following code to display the document:

```
# Document
documentName = "benefit-app.png"
display(Image(filename=documentName))
```

You can check the following sample filled-in benefit application form:

Application for Benefits

Division of Family and Children Services

(Complete this application and return it to your LOCAL COUNTY DFCS office.)

What Am I Applying For: (Check all that apply)

❑ **Food Stamps (Supplemental Nutrition Assistance Program (SNAP)**
The Supplemental Nutrition Assistance Program (SNAP), formerly known as Food Stamps, is a federally funded program that provides monthly benefits to low-income households to help pay for the cost of food. The program also provides nutrition education to families to meet their food and nutritional needs and employment and training opportunities to help families gain employment that leads to less dependence on SNAP.

❑ **Temporary Assistance for Needy Families (TANF)**
Temporary Assistance for Needy Families (TANF) provides temporary monthly cash payments, single cash payments, or other support services, to strengthen eligible families with children. If you are the child's parent, or the caretaker who would like to be included in the grant, we will require you to participate in a work program.
    ❑ **Grandparents Raising Grandchildren (GRG)**
    Grandparents Raising Grandchildren (GRG) will provide additional cash payments so that children can be cared for in the homes of their grandparents. Applicants must apply for TANF to be eligible for GRG.

❑ **Refugee Cash Assistance**
The Refugee Cash Assistance program provides financial assistance to refugee households who are not eligible for the TANF program. The term refugee includes refugees, Cuban/ Haitian Entrants, victims of human trafficking, Amerasians, and unaccompanied refugee minors.

❑ **Medicaid**
Medicaid offers medical coverage to elderly, blind or disabled adults, pregnant women, children, and families. When you apply, we will look at all Medicaid programs and decide which ones you may be eligible to receive.

Please fill out the chart below about the applicant.

First Name	Middle Initial	Last Name	Suffix
Jane		Doe	

Street Address Where You Live		Apt	
Any Street			

City	State	Zip Code	
Any City	Any State	12345	

Mailing Address (If different)

Main Telephone Number	Other Contact Number	Email Address (Optional)

E-mail Communication Yes___ or No___ (optional)	Texting: Yes__ or No__ (optional)

What is your Preferred Language? English	If an interview is required, will you need an interpreter? Yes___ or No ____

Figure 9.2 – Sample benefit application form

5. After we get the benefit application, we are using Amazon Textract to accurately extract all the elements from the document. We will be getting Textract boto3 client object as `client`. Here, we are showing you another way to get the Textract botot3 object, by specifying an endpoint and region.

6.   We are also processing the document as a local image instead of storing it in a AWS S3 bucket, we are using the `analyze_document()` Textract Sync Form API to process the document. This API takes blob of byte, so we are extracting raw bytes from the document and passing it as a blob to the API. As this document is a form type of document, we are using the Textract API with `FeatureType` set to FORM. We have defined this code in a `calltextract()` method for further reuse:

```
client = boto3.client(service_name='textract',
         region_name= 'us-east-1',
         endpoint_url='https://textract.us-east-1.
amazonaws.com')


#Alternative way to call Tetxract client with API.
# process using image bytes

def calltextract(documentName):
    with open(documentName, 'rb') as file:
            img_test = file.read()
            bytes_test = bytearray(img_test)
            print('Image loaded', documentName)
    # process using image bytes
    response = client.analyze_document(Document={'Bytes':
bytes_test}, FeatureTypes=['FORMS'])
    return response
```

7.   Let's call the preceding-defined method `calltextract()` to parse the document and print the key-value response:

```
response= calltextract(documentName)
print(response)
```

You can check the following sample Textract form response:

Image loaded benefit-app.png
{'DocumentMetadata': {'Pages': 1}, 'Blocks': [{'BlockType': 'PAGE', 'Geometry': {'BoundingBox': {'Width': 1.0, 'Hei
ght': 1.0, 'Left': 0.0, 'Top': 0.0}, 'Polygon': [{'X': 1.5696498570978468e-16, 'Y': 0.0}, {'X': 1.0, 'Y': 9.5547400
94542871e-17}, {'X': 1.0, 'Y': 1.0}, {'X': 0.0, 'Y': 1.0}]}, 'Id': '41136fb3-0176-4d77-9ca6-0b7722b56c1e', 'Relatio
nships': [{'Type': 'CHILD', 'Ids': ['c507c612-a765-421e-b0f1-80181890adf2', '68539ddf-c73d-46d5-81a3-6d0b34eae086',
'29abb896-46a4-46a3-a036-a5944638c0a9', 'daada7df-dfbb-4ce2-bd00-643a7201982b', '7b405fd7-a0bb-4c9c-a3d5-f5f426155d
c1', '15c7386c-83a4-425f-83aa-90b447f0b602', '686c10f3-81ef-47a9-ab67-471766659025', '23839ba2-a5bd-40b8-872d-2a11c
a83c880', 'af90d3e7-47eb-4ed7-8297-4a74286f84c4', 'e1308f76-78bd-4eed-b5b2-a2d94e516343', 'd306a6f1-891a-46fb-a7e9-
81bb79442824', '8937c36a-15ad-486c-8303-4f5f0002712a', 'fd79d1d0-94e7-4682-8c5b-e6b6111ed5b8', '87a97022-0059-46b3-
9499-5e0da41154f7', 'b8dbb28f-912f-487e-803b-e87e9a28f3ac', 'e57216e3-b566-41f1-b422-f83df44e82fc', '5a058c39-c227-
4ba5-812e-cba867b936fd', '778ee7e0-1b64-4df1-a1bf-beb8553e8976', '7abae22c-6671-458c-959a-e2a108ef0cf8', '81e92882-

Figure 9.3 – Textract response for the benefit application form

8. Once we have the raw Textract JSON response, we are parsing the JSON response to get the key value from the JSON data with the following code. This JSON response parser is defined in a getformkeyvalue() method for further reuse:

```
#Extract key values
# Iterate over elements in the document
from trp import Document
def getformkeyvalue(response):
    doc = Document(response)
    key_map = {}
    for page in doc.pages:
        # Print fields
        for field in page.form.fields:
            if field is None or field.key is None or
field.value is None:
                continue
            key_map[field.key.text] = field.value.text
    return key_map
```

9. Let's now call the previously defined method, getformkeyvalue(), to get parsed key-value results from the document:

```
get_form_keys = getformkeyvalue(response)
print(get_form_keys)
```

You can check the sample key-value output extracted accurately from the document, such as First Name, City, Street Address Where You Live, What is your Preferred Language?, from the benefit application document:

```
{'City': 'Any City', 'First Name': 'Jane', 'Zip Code': '12345', 'Street Address Where You Live': 'Any Street', 'Sta
te': 'Any State', 'Braille': ';', 'What is your Preferred Language?': 'English', 'Yes': 'NOT_SELECTED', 'Food Stamp
s (Supplemental Nutrition Assistance Program (SNAP) The Supplemental Nutrition Assistance Program (SNAP), formerly
known as Food Stamps, is a federally funded program that provides monthly benefits to low-income households to help
pay for the cost of food. The program also provides nutrition education to families to meet their food and nutritio
nal needs and employment and training opportunities to help families gain employment that leads to less dependence
SNAP.': 'NOT_SELECTED', 'Refugee Cash Assistance The Refugee Cash Assistance program provides financial assistance
to refugee households who are not eligible for the TANF program. The term refugee includes refugees, Cuban/Haitian
Entrants, victims of human trafficking, Amerasians, and unaccompanied refugee minors.': 'NOT_SELECTED', 'Last Name
': 'Doe', "Temporary Assistance for Needy Families (TANF) Temporary Assistance for Needy Families (TANF) provides t
```

Figure 9.4 – Textract Queries response for the benefit application form

In the preceding steps, you saw how to use Textract's form APIs and parser code to get key-values from the benefit application form. Now, we will show you another way to extract information from the benefit application form. For this code, we will show you how to extract What is your Preferred

Language? from the same benefit application document with Amazon Textract Queries. To use Textract Queries, we are sending the What is the Preferred Language? question to the Textract analyze_document() API and getting the response:

```
response = None
with open(documentName, 'rb') as document:
    imageBytes = bytearray(document.read())
    # Call Textract
    response = client.analyze_document(
        Document={'Bytes': imageBytes},
        # new QUERIES Feature Type for Textract Queries
        # We could add additional Feature Types like FORMS and/
or TABLES
        # FeatureTypes=["QUERIES", "FORMS", "TABLES"],
        FeatureTypes=["QUERIES"],
        QueriesConfig={
            "Queries": [{
                "Text": "What is the prefered language?",
                "Alias": "LANGUAGE"
            }
        ]
        })
    print(response)
```

We are getting English as the preferred language extracted from the sample benefit document.

In the preceding code sample, we saw how to use the Amazon Textract FORM API to extract all information from the benefit application form.

As discussed before, for benefit enrollment, the user has to submit additional supporting documents to the benefit application form. We will use one such document, of the SSN card type, for our example code:

1. Let's first display our SSN card document with the following code sample:

    ```
    # Document
    documentName = "ssn_Jesse_123456789.jpg"
    display(Image(filename=documentName))
    ```

You can see the sample SSN card as follows:

Figure 9.5 – Sample SSN document

2. We are interested in extracting the SSN and name from the SSN card. To extract the SSN and name, we are using Amazon Textract's Queries feature. We are passing questions such as What is ssn number? and What is the name? to Amazon Textract's analyze_document() method:

```
response = None
with open(documentName, 'rb') as document:
    imageBytes = bytearray(document.read())
    # Call Textract
    response = client.analyze_document(
        Document={'Bytes': imageBytes},
        # new QUERIES Feature Type for Textract Queries
        # We could add additional Feature Types like
FORMS and/or TABLES
        # FeatureTypes=["QUERIES", "FORMS", "TABLS"],
        FeatureTypes=["QUERIES"],
        QueriesConfig={
            "Queries": [{
                "Text": "What is the ssn number?",
                "Alias": "SSN"
            },
```

```
                        {
                            "Text": "What is the name?",
                            "Alias": "NAME"
                        }
                    ]
                })
            print(response)
```

3. You can see Amazon Textract has accurately extracted the SSN and name from the SSN card document.

We discussed the implementation of the benefit enrollment application with sample code. Now, we will dive into another use case, claims processing in the insurance industry.

Understanding insurance claims processing extraction with IDP

Claims processing is the process to check for the validity and authenticity of a claim to verify the right financial responsibility. Most often, this process is manual, which is time consuming, error prone, and expensive. Also, you'll need skilled professionals for the verification of data. We will show you how IDP AI services can help in automating a claims processing use case. Let's now dive into the technical details:

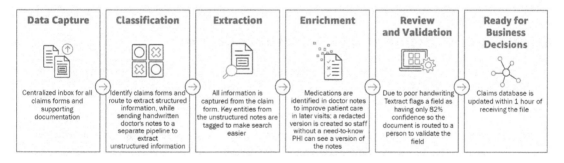

Figure 9.6 – IDP flow for insurance claims processing

The data capture and document classification stages of the IDP pipeline

We are using Amazon S3 to store our claims along with supporting documents, such as a claims document (CMS-1500 or UB04), ID document, invoice, or providers note. After these documents are stored, we need to accurately classify them before we apply extraction logic and further processing on these documents. For example, we want to extract the name and ID number from the CMS application form, but want to extract amount details from the invoice document. Let's look into the classification training architecture using AWS AI services:

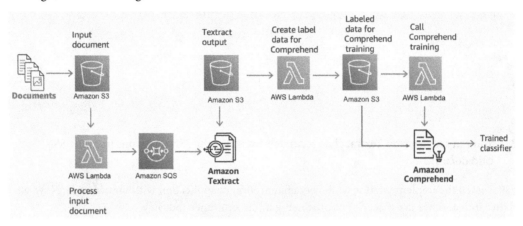

Figure 9.7 – Architecture for Comprehend training

In this architecture, we store all input documents in an Amazon S3 bucket, then in an event-driven manner, we call Amazon Textract to extract these documents. We then use AWS Lambda to create labeled training data in a `.csv` file for Comprehend model training. Once we have the required labeled data, we call Amazon Comprehend custom classification to train the Comprehend model. Once the training is done, the Amazon Comprehend custom classification model gets stored in an Amazon S3 bucket. Then, we have the option to create a real-time endpoint or run an analysis job for inference. I recommend checking the details on training the Comprehend classifier in *Chapter 2, Document Capture and Categorization*.

Document extraction stage of the IDP pipeline

During the document extraction stage of the IDP pipeline for claims processing, we are using Amazon Tetxract for the accurate extraction of elements from the CMS-1500 form. Let's work through the implementation as follows:

1. First, let's check our claims form. We are using a stripped-down version of our claims form. You can run the following command to check the image:

    ```
    # Document
    documentName = "cms1500-1.png"
    display(Image(filename=documentName))
    ```

You can check the following figure:

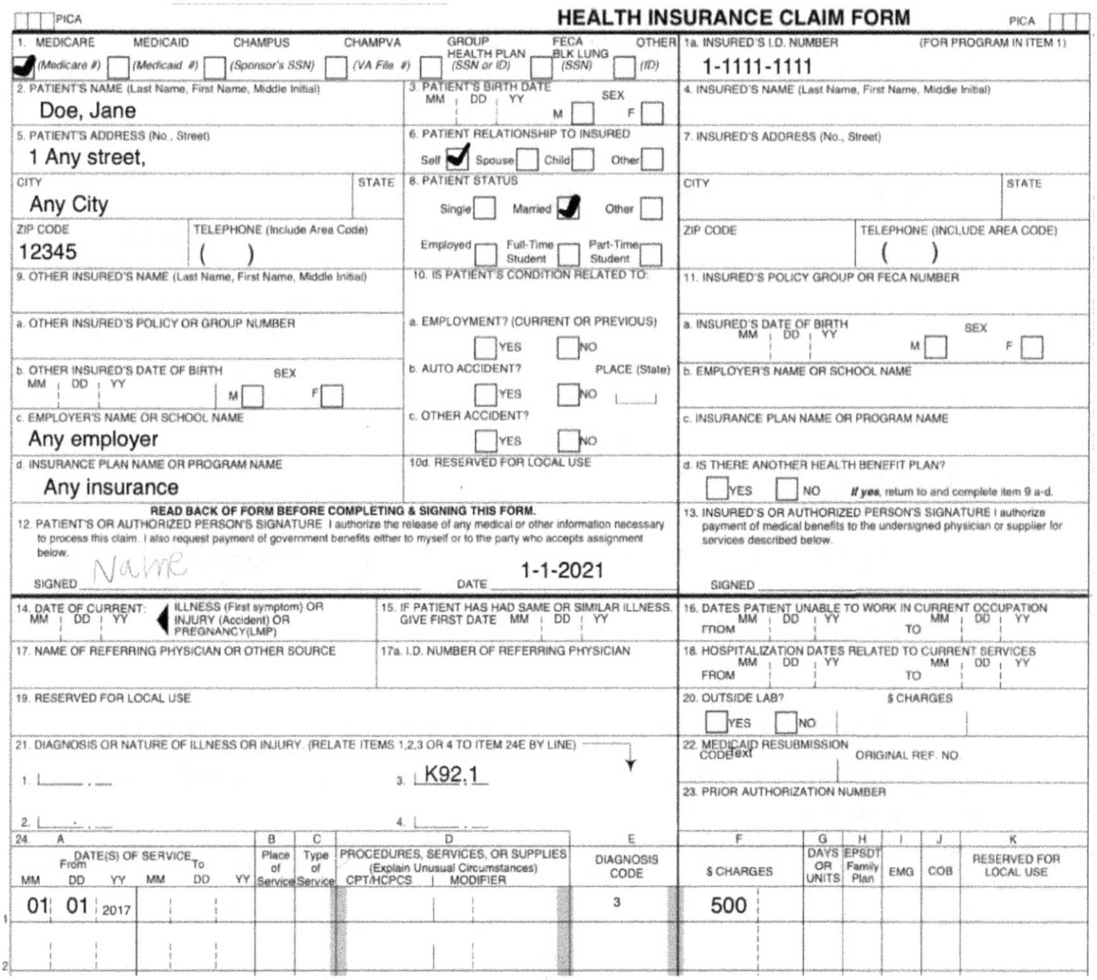

PICA HEALTH INSURANCE CLAIM FORM PICA

1. MEDICARE (Medicare #) ✔ MEDICAID (Medicaid #) CHAMPUS (Sponsor's SSN) CHAMPVA (VA File #) GROUP HEALTH PLAN (SSN or ID) FECA BLK LUNG (SSN) OTHER (ID)
1a. INSURED'S I.D. NUMBER (FOR PROGRAM IN ITEM 1): 1-1111-1111
2. PATIENT'S NAME (Last Name, First Name, Middle Initial): Doe, Jane
3. PATIENT'S BIRTH DATE MM DD YY SEX M F
4. INSURED'S NAME (Last Name, First Name, Middle Initial)
5. PATIENT'S ADDRESS (No., Street): 1 Any street,
6. PATIENT RELATIONSHIP TO INSURED: Self ✔ Spouse Child Other
7. INSURED'S ADDRESS (No., Street)
CITY: Any City STATE
8. PATIENT STATUS: Single Married ✔ Other; Employed Full-Time Student Part-Time Student
CITY STATE
ZIP CODE: 12345 TELEPHONE (Include Area Code) ( )
ZIP CODE TELEPHONE (INCLUDE AREA CODE) ( )
9. OTHER INSURED'S NAME (Last Name, First Name, Middle Initial)
10. IS PATIENT'S CONDITION RELATED TO:
11. INSURED'S POLICY GROUP OR FECA NUMBER
a. OTHER INSURED'S POLICY OR GROUP NUMBER
a. EMPLOYMENT? (CURRENT OR PREVIOUS) YES NO
a. INSURED'S DATE OF BIRTH MM DD YY SEX M F
b. OTHER INSURED'S DATE OF BIRTH MM DD YY SEX M F
b. AUTO ACCIDENT? YES NO PLACE (State)
b. EMPLOYER'S NAME OR SCHOOL NAME
c. EMPLOYER'S NAME OR SCHOOL NAME: Any employer
c. OTHER ACCIDENT? YES NO
c. INSURANCE PLAN NAME OR PROGRAM NAME
d. INSURANCE PLAN NAME OR PROGRAM NAME: Any insurance
10d. RESERVED FOR LOCAL USE
d. IS THERE ANOTHER HEALTH BENEFIT PLAN? YES NO *If yes*, return to and complete item 9 a-d.
READ BACK OF FORM BEFORE COMPLETING & SIGNING THIS FORM.
12. PATIENT'S OR AUTHORIZED PERSON'S SIGNATURE I authorize the release of any medical or other information necessary to process this claim. I also request payment of government benefits either to myself or to the party who accepts assignment below.
SIGNED Name DATE 1-1-2021
13. INSURED'S OR AUTHORIZED PERSON'S SIGNATURE I authorize payment of medical benefits to the undersigned physician or supplier for services described below.
SIGNED
14. DATE OF CURRENT: MM DD YY ILLNESS (First symptom) OR INJURY (Accident) OR PREGNANCY(LMP)
15. IF PATIENT HAS HAD SAME OR SIMILAR ILLNESS. GIVE FIRST DATE MM DD YY
16. DATES PATIENT UNABLE TO WORK IN CURRENT OCCUPATION FROM MM DD YY TO MM DD YY
17. NAME OF REFERRING PHYSICIAN OR OTHER SOURCE
17a. I.D. NUMBER OF REFERRING PHYSICIAN
18. HOSPITALIZATION DATES RELATED TO CURRENT SERVICES FROM MM DD YY TO MM DD YY
19. RESERVED FOR LOCAL USE
20. OUTSIDE LAB? YES NO $ CHARGES
21. DIAGNOSIS OR NATURE OF ILLNESS OR INJURY. (RELATE ITEMS 1,2,3 OR 4 TO ITEM 24E BY LINE)
1. 2. 3. K92.1 4.
22. MEDICAID RESUBMISSION CODE ORIGINAL REF. NO.
23. PRIOR AUTHORIZATION NUMBER

24. A DATE(S) OF SERVICE From MM DD YY To MM DD YY	B Place of Service	C Type of Service	D PROCEDURES, SERVICES, OR SUPPLIES (Explain Unusual Circumstances) CPT/HCPCS MODIFIER	E DIAGNOSIS CODE	F $ CHARGES	G DAYS OR UNITS	H EPSDT Family Plan	I EMG	J COB	K RESERVED FOR LOCAL USE
1 01 01 2017				3	500					
2										

Figure 9.8 – Sample claims CMS application

2. Now, we are defining a `calltextract()` method. In this method, we get a `boto3` client of Amazon Textract. We are calling the `analyze_document()` API of Textract to get form API information from the document. We are passing in the document from a local file. Thus, we are sending all the raw bytes to the `analyze_document()` API:

```
# process using image bytes
def calltextract(documentName):
    client = boto3.client(service_name='textract',
         region_name= 'us-east-1',
```

```
                       endpoint_url='https://textract.us-east-1.
        amazonaws.com')
            with open(documentName, 'rb') as file:
                    img_test = file.read()
                    bytes_test = bytearray(img_test)
                    print('Image loaded', documentName)
            # process using image bytes
            response = client.analyze_document(Document={'Bytes':
        bytes_test}, FeatureTypes=['FORMS'])
            return response
```

3. Now, call the previously defined method, `calltextract()`, to extract the CMS document accurately:

```
response= calltextract(documentName)
print(response)
```

4. We extracted all key-value pairs from the CMS form. Now, let's parse the response and print it. We are calling the `getformkeyvalue()` method as defined in step 10 of the *Automating benefit enrollment process with IDP* section:

```
get_form_keys = getformkeyvalue(response)
print(get_form_keys)
```

You can check all key-value responses, as follows. You can see **Name**, **Insured ID**, **Insurance Plan**, and more have been extracted accurately:

```
{'YES': 'NOT_SELECTED', 'NO': 'NOT_SELECTED', 'ZIP CODE': '12345', 'Full-Time Student': 'NOT_SELECTED', 'CITY': 'An
y City', '5. PATIENT'S ADDRESS (No., Street)": '1 Any street,', 'Part-Time Student': 'NOT_SELECTED', "c. EMPLOYER'S
NAME OR SCHOOL NAME": 'Any employer', "2. PATIENT'S NAME (Last Name, First Name, Middle Initial)": 'Doe, Jane', 'Ot
her': 'NOT_SELECTED', 'Single': 'NOT_SELECTED', 'TELEPHONE (Include Area Code)': '( )', 'M': 'NOT_SELECTED', 'TELEP
HONE (INCLUDE AREA CODE)': '( )', 'Self': 'SELECTED', 'Child': 'NOT_SELECTED', 'd. INSURANCE PLAN NAME OR PROGRAM N
AME': 'Any insurance', 'Spouse': 'NOT_SELECTED', "(Sponsor's SSN)": 'NOT_SELECTED', 'DATE': '1-1-2021', '(Medicaid
#)': 'NOT_SELECTED', 'F': 'NOT_SELECTED', 'Married': 'SELECTED', '(Medicare #)': 'SELECTED', '(ID)': 'NOT_SELECTED
', '3.': 'K92.1', 'SIGNED': 'Name', 'FECA BLK LUNG (SSN)': 'NOT_SELECTED', "1a. INSURED'S I.D. NUMBER": '1-1111-111
1', '(VA File #)': 'NOT_SELECTED'}
```

Figure 9.9 – Textract response for CMS application

For claims processing, most often additional supporting documents, such as a doctor's note, discharge summary, invoice, and insurance ID, are submitted. Or, for different types of insurance claim use cases, document types may vary and you might have, for example, an ACORD form or APS document. To accurately extract these documents, we will follow similar implementation steps as mentioned in this section. I'd recommend you check *Chapter 3, Accurate Document Extraction with Amazon Textract*, and *Chapter 4, Accurate Extraction with Amazon Comprehend*, to use the right extraction API for your document type.

Understanding insurance claims processing document enrichment and review and verification

Let's now check how the review and verification stage for our insurance claims processing use case works. We mentioned the review and verification steps in detail in *Chapter 6*, *Review and Verification of Intelligent Document Processing*. If you have not gone through that chapter, I recommend going through it first. In *Chapter 6*, *Review and Verification of Intelligent Document Processing*, we described high-level types of document verification steps. For example, we have a completeness check and an accuracy check to process a document in the review and validation stage of the IDP pipeline. During this stage, the user defines business rules to be checked against the extracted elements from documents. Let's now see an example implementation to check for completeness of the document.

If you have not already executed step 1 to step 6 of the *Understanding insurance claims processing extraction with IDP* section, we recommend executing those steps before diving deep into the following solution.

We are defining two business rules here:

- We need to ensure that the ZIP code is not empty and is a numerical value

- The Insurance ID should be an 11-digit number

Now, let's check the sample code to do this validation:

1. We defined a `validate()` method, which takes in a parsed JSON response from Amazon Textract and checks for the preceding two mentioned business conditions, such as checking for the ZIP code and insurance ID:

```python
def validate(body):
    json_acceptable_string = body.replace("'", "\"")
    json_data = json.loads(body)
    zip = json_data['ZIP CODE']
    id = json_data['1a. INSURED\'S I.D. NUMBER']
    diagnosiscode = json_data["3."]
    print(diagnosiscode)
    print(id)
    print(zip)
    if(not zip.strip().isdigit()):
        return False, id, diagnosiscode, "Zip code
invalid"
    length = len(id.strip())
    if(length != 11):
```

```
                return False, id, diagnosiscode, "Invalid claim
     Id"
          return True, id, diagnosiscode, "Ok"
```

2. We call the `validate()` method defined in step 1 and print the response:

```
# Validate
textract_json= json.dumps(get_form_keys,indent=2)
res, formid, diagnosiscode, result = validate(textract_
json)
print(result)
print(formid)
```

3. For this sample application form, the insurance ID and ZIP code are extracted accurately and validation is successful.

We can extend this to do additional validation, such as *medical coding*-level validation for healthcare claims processing. Most often for healthcare claims processing, additional clinical data is requested from the healthcare provider to check for the validity of the claims. For this example, we will infer the ICD10-CM code from provider notes with Amazon Comprehend Medical and compare the code mentioned against the ICD10-CM code mentioned in our CMS application form. During the verification process, the reviewer checks for accurate medical coding mentioned in the application form with a clinical record. This is an example of cross-document validation. Let's now dive into the implementation details for the cross-document validation process:

1. First, let's check the provider note. You can run the following command to check the image:

```
documentName = "doctornotes1.png"
display(Image(filename=documentName))
```

A sample provider note is shown here:

The patient is an 86-year-old female admitted for evaluation of abdominal pain and bloody stools. The patient has colitis and also diverticulitis, undergoing treatment. During the hospitalization, the patient complains of shortness of breath, which is worsening. The patient underwent an echocardiogram, which shows severe mitral regurgitation and also large pleural effusion. This consultation is for further evaluation in this regard. As per the patient, she is an 86-year-old female, has limited activity level. She has been having shortness of breath for many years. She also was told that she has a heart murmur, which was not followed through on a regular basis.

Figure 9.10 – Provider note

2. If you have not already executed step 1 to step 6 of the *Understanding insurance claims processing extraction with IDP* section, we recommend executing those steps before diving deep, then using the following code to print the diagnosis code:

```
print("diagnosis code from claim form", diagnosiscode)
print("diagnosis code from raw doctor's note", code)
```

In the following figure, you can see the sample output:

```
diagnosis code from claim form K92.1
diagnosis code from raw doctor's note K92.1
```

Figure 9.11– Medical coding extraction

You have now learned how to use the IDP review and verification stage to accurately check for the completeness with medical coding level cross-document verification. Using the preceding implementation code, we successfully checked the validity of a claim, and it was a successful validation. At times, we receive an application form with incomplete information; can we process those forms in an automated way using the AWS IDP pipeline to carry out a completeness check? Let's now check out an example of an application form like this with incomplete information and how we can process it in an automated way with IDP.

Claims processing for an invalid claims form

For this example, we are using a claims form where the expected number of digits for the insurance ID is 11, but we'll intentionally fill it in with an ID with a different number of digits. Let's check how we can use the AWS IDP services to automatically review this claim document:

1. First, let's check the invalid claim form. You can run the following command to check the image:

```
InvalidDocument = "failedtest.png"
display(Image(filename=InvalidDocument))
```

A sample claims form is shown here:

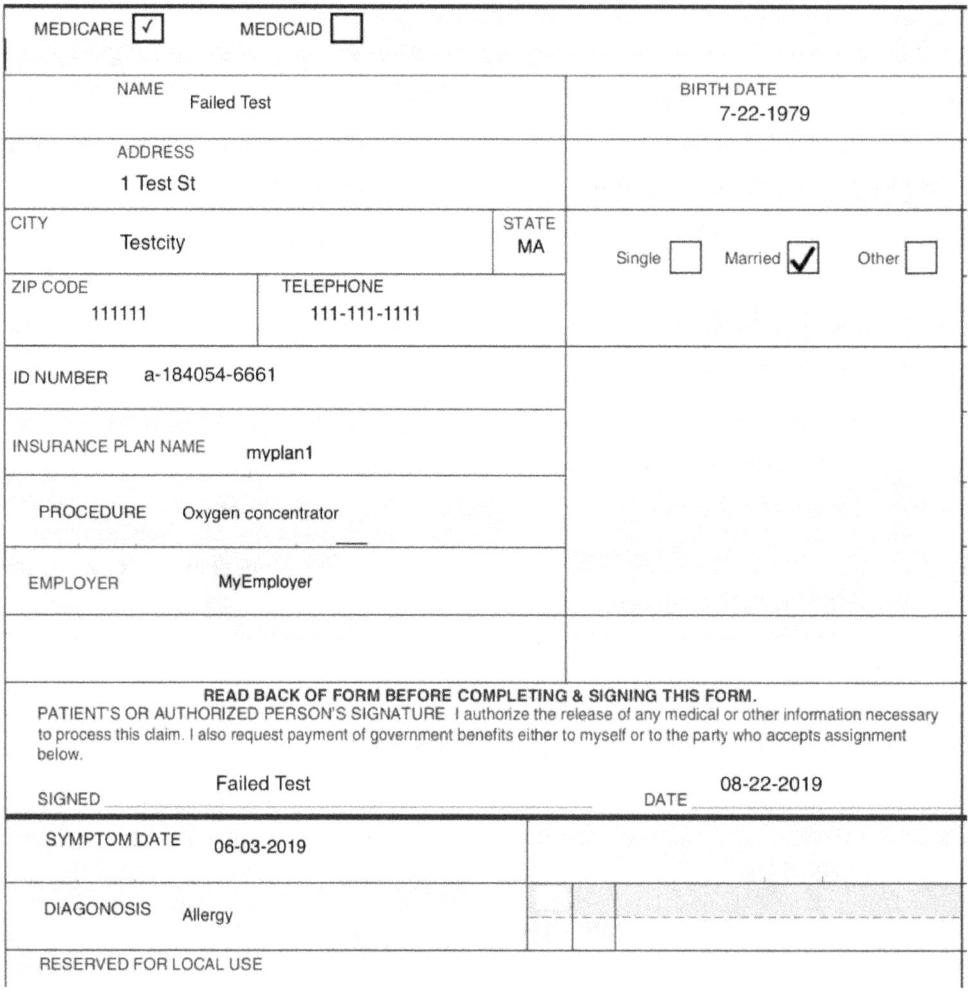

MEDICARE ☑ MEDICAID ☐

NAME Failed Test

BIRTH DATE 7-22-1979

ADDRESS 1 Test St

CITY Testcity

STATE MA

Single ☐ Married ☑ Other ☐

ZIP CODE 111111

TELEPHONE 111-111-1111

ID NUMBER a-184054-6661

INSURANCE PLAN NAME myplan1

PROCEDURE Oxygen concentrator

EMPLOYER MyEmployer

READ BACK OF FORM BEFORE COMPLETING & SIGNING THIS FORM.
PATIENT'S OR AUTHORIZED PERSON'S SIGNATURE I authorize the release of any medical or other information necessary to process this claim. I also request payment of government benefits either to myself or to the party who accepts assignment below.

SIGNED Failed Test DATE 08-22-2019

SYMPTOM DATE 06-03-2019

DIAGONOSIS Allergy

RESERVED FOR LOCAL USE

Figure 9.12 – Sample invalid CMS application

2. Call the `calltextract()` method defined in *step 4* of the *Understanding Insurance claims processing Extraction with IDP* section and pass in the invalid claims form:

```
response = calltextract(InvalidDocument)
```

3. We extracted all key-value pairs from the invalid CMS form. Now let's parse the response and print it. We are calling the `getformkeyvalue()` method as defined in step 10 of the *Automating the benefit enrollment process with IDP* section:

```
get_form_keys = getformkeyvalue(response)
print(get_form_keys)
```

Let's check the key-value pairs, as shown:

```
{'Married': 'SELECTED', 'Other': 'NOT_SELECTED', 'MEDICAID': 'NOT_SELECTED', 'Single': 'NOT_SELECTED', 'TELEPHONE':
'111-111-1111', 'ZIP CODE': '111111', 'STATE': 'MA', 'INSURANCE PLAN NAME': 'myplan1', 'ID NUMBER': 'a-184054-6661
', 'ADDRESS': '1 Test St', 'NAME': 'Failed Test', 'DATE': '08-22-2019', 'SIGNED': 'Failed Test', 'MEDICARE': 'SELEC
TED', 'DIAGONOSIS': 'Allergy', 'BIRTH DATE': '7-22-1979', 'CITY': 'Testcity', 'SYMPTOMDATE': '06-03-2019', 'PROCEDU
RE': 'Oxygen concentrator', 'EMPLOYER': 'MyEmployer'}
```

Figure 9.13 – Textract response invalid CMS application

4. With the following code, we are checking for the validity of the document and whether the number of digits is 11 for the insurance ID. But the insurance ID extracted from the document fails the business rule, and we get an `"Invalid claim Id"` response, as follows:

```
#In Validate
textract_json= json.dumps(get_form_keys,indent=2)
json_data = json.loads(textract_json)
id = json_data['ID NUMBER']
print(id)
length = len(id.strip())
if(length != 11):
    print("Invalid claim Id")
```

At times, we have a requirement to include human beings to review and correct the elements during the claims processing use case. I'd recommend checking the *Learning Document review process with human loop* section in *Chapter 6*, *Review and Verification of Intelligent Document Processing*, to see how to include a human being in an automated manner to review documents as needed. In this chapter, we checked how we can use IDP services to verify claims and the supporting documents for automatic validation.

Summary

In this chapter, we discussed the IDP pipeline and how this can apply to insurance industry use cases. We discussed the benefit enrollment application and how we can use AWS AI services to automatically check for benefit eligibility before we finalize the benefit application process.

We then dove deep into an insurance claims processing use case. We also discussed how to automate the extraction and enrichment process of a claims document and its supporting documents with AWS AI services. We also defined business rules and checked the validity of our claims against the defined business rules. Finally, we dove deep into the implementation details with AWS IDP services to auto-adjudicate a claim. We also used an invalid claim document to check for its completeness.

In the next chapter, we will extend IDP to an additional industry use case. We will dive deep into a mortgage application processing use case and see how IDP can help automate it.

10

Intelligent Document Processing – Mortgage Processing

In the previous chapter, you understood the challenges involved when processing documents in the insurance industry. We looked into the pain points when processing documents to derive meaningful insights at scale. We also discussed potential solutions using AWS AI services such as Amazon Textract and Amazon Comprehend to automate this **Intelligent Document Processing** (**IDP**) pipeline. We dove deep into the solutions to process benefit enrollment applications; further, we discussed the automation of document processing for claims processing use cases. We will now change gear and look into use cases in the financial industry to process documents for mortgage loan application processing. We will also dive into the IDP pipeline and how IDP can help to automate financial industry use cases following the right security and compliance requirements. We will also discuss additional financial industry use cases around **Know Your Customer** (**KYC**) standards. We will navigate through the following sections in this chapter:

- Automating mortgage processing data capture and data categorization with IDP
- Automating mortgage processing data capture and data categorization with IDP
- Understanding financial industry use cases for document processing

Technical requirements

For this chapter, you will need access to an AWS account. Before getting started, we recommend that you create an AWS account by referring to the AWS account setup and Jupyter notebook creation steps as mentioned in the *Technical requirements* section in *Chapter 2, Document Capture and Categorization*.

You can find Chapter-9 code sample in GitHub: `https://github.com/PacktPublishing/Intelligent-Document-Processing-with-AWS-AI-ML-/tree/main/chapter-10`. Also recommend to check availability of AI service in your AWS regions before using it.

Automating mortgage processing data capture and data categorization with IDP

Before diving deep, let's first understand how mortgage companies and mortgage loan application processing work.

Anyone that has bought a house has most likely gone through the mortgage process. It is the time between applying for a home loan and closing the process. During the process, the lender verifies and confirms the buyer's credit record, and checks for the accuracy and completeness of records before approving the loan. **Mortgage processing** consists of multiple stages, such as the collection of all documents, loan application processing, underwriting, and closing. During the process, a loan processor or underwriter assembles all the required documents, then administers and processes the loan application paperwork. The mortgage processor, or loan processor, collects all the required documents and guides the buyer in finalizing all the paperwork. The steps can be quite complex, but the loan officer helps you find the right loan for your budget and needs. This loan processor collects your financial documents and analyzes your credit report. Once all the required documents are collected and verified, the loan processor passes them to the underwriter. Following up on the process, the underwriter analyzes and gets insights from the loan application documents.

Organizations in the lending and mortgage industry process tons of documents. Whether you are talking about new mortgage applications or mortgage refinancing, the lending industry has to deal with tons of documents per application. Moreover, the number of application processes at scale adds to the challenge of document processing. This manual document processing is expensive and time-consuming. Legacy document extraction tools, such as **optical character recognition** (**OCR**) tools, can extract information from documents but most often, they just extract a list of words, which makes it difficult to derive insights from the documents. IDP with AWS AI services helps to automate mortgage processing and reduce costs. The following figure shows the high-level solutions architecture for document processing in the lending industry:

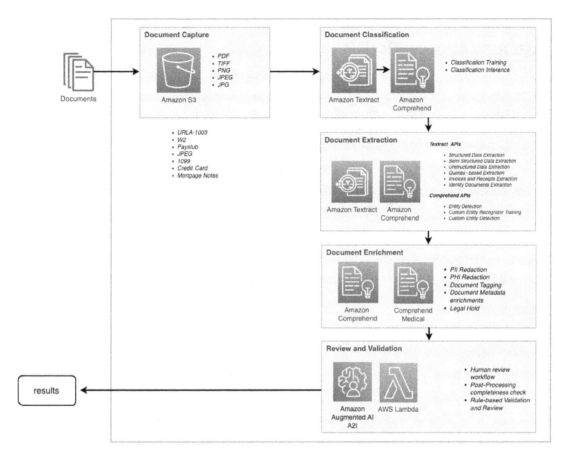

Figure 10.1 – Flow architecture – IDP – financial industry

All documents get stored and aggregated in a secure data store in Amazon S3 in the data capture stage of IDP. Once documents get stored in S3, we categorize these documents during the document classification stage for further document extraction. In the document extraction stage, we extract all elements from the financial and supporting documents for mortgage processing. Most often, for document processing, we do enrichment such as redacting any sensitive information from documents in the document enrichment stage of IDP. The critical component of mortgage loan application processing is to verify and validate documents for accuracy and completeness. During the review and verification stage of IDP, the loan processor or underwriter defines the business rules and checks the information for validity, authenticity, and completeness of the mortgage loan application.

Now let's dive into each stage of the mortgage loan application with IDP. We will look into an automation solution to automate the loan application process with AWS IDP services such as Amazon Textract and Amazon Comprehend.

Automating mortgage processing data capture and data categorization with IDP

The first stage of the IDP pipeline is the **data capture stage**. During this stage, all documents (such as URLA-1003, Form W-2, pay stubs, bank statements, credit card statements, mortgage notes, Form 1099, ID documents such as a passport and driver's license, and any other documents) are collected and aggregated in a central secure data store on Amazon S3. You can define the right access control for the data on S3. This is the data capture stage of IDP.

At times, we know the document type, and can do further extraction. But most often, we do not have any specific way of identifying the documents; in that scenario, we need to *classify documents* before further extraction. We can use Textract to extract raw text from any type of document. Then, we can create sample label data for training a Comprehend classifier. Amazon Comprehend classification can help accurately categorize documents for mortgage application processing. In the following diagram, you can see input as a random set of documents. They pass through Amazon Textract for extraction, then we create sample labeled data for Amazon Comprehend classification training. Once the Amazon Comprehend classifier is trained, we deploy the endpoint and can get inference results to categorize documents.

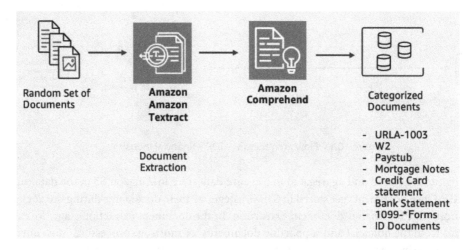

Figure 10.2 – Document classification – financial industry use case

At times, we receive documents joined in a single PDF. We can use document splitter solutions with Amazon Textract and Amazon Comprehend to split documents into categories. To build your document classifier with the AWS IDP solution, I recommend checking *Chapter 2, Document Capture and Categorization*. To check the solution of the document splitter, please check the *references* section.

Understanding mortgage processing extraction and enrichment with IDP

As discussed before, during mortgage loan application processing, we check for the validity, identity, and authenticity of the mortgage loan requester. Most often, this process is manual, which is time-consuming, error-prone, and expensive. We will show you how IDP AI services can help to automate mortgage processing use cases. Let's now dive into the technical details.

For complete code walk-through follow the steps in the Notebook: `https://github.com/ PacktPublishing/Intelligent-Document-Processing-with-AWS-AI-ML-/ blob/main/chapter-10/mortgage-processing-10.ipynb`:

1. First, let's import the required libraries to run the sample code:

    ```
    import boto3
    import json
    import uuid
    import io
    from io import BytesIO
    import sys
    from pprint import pprint
    from IPython.display import Image, display
    from PIL import Image as PImage, ImageDraw
    ```

2. In the loan application package, we may receive a list of documents, such as the URLA-1003 form. In the following code block, we are creating a `textract` client and calling the `analyze_document()` API to extract all elements from the URLA-1003 form. This document is an example of a semi-structured document type with key-value elements. Thus, we are calling the `analyze_document()` API with FORMS as FeatureTypes. We are processing Document as a local file. To process this from a local file instead of uploading to the Amazon S3 bucket, we are passing in raw bytes from the image. We have defined the `calltextract()` method for accurate extraction with Amazon Textract:

    ```
    # Document
    documentName = "URLA-1003.pdf"
    client = boto3.client(service_name='textract',
            region_name= 'us-east-1',
            boto3.client('textract')
    # process using image bytes
    def calltextract(documentName):
        with open(documentName, 'rb') as file:
    ```

```
            img_test = file.read()
            bytes_test = bytearray(img_test)
            print('Image loaded', documentName)
    # process using image bytes
    response = client.analyze_document(Document={'Bytes':
bytes_test}, FeatureTypes=['FORMS'])
    return response
```

3. Now, let's call the `calltextract()` method we just defined and check the response:

```
response= calltextract(documentName)
print(response)
```

4. Next, you can see a snippet of the response from Amazon Textract and the JSON response of the extracted elements:

Image loaded URLA-1003.pdf
{'DocumentMetadata': {'Pages': 1}, 'Blocks': [{'BlockType': 'PAGE', 'Geometry': {'BoundingBox': {'Width': 1.0, 'Hei
ght': 1.0, 'Left': 0.0, 'Top': 0.0}, 'Polygon': [{'X': 0.0, 'Y': 2.3413836913732666e-07}, {'X': 1.0, 'Y': 0.0}, {'X
': 1.0, 'Y': 1.0}, {'X': 4.185015995972208e-07, 'Y': 1.0}]}, 'Id': 'f724a516-178c-4247-aabd-0a2d0d2d5a9f', 'Relatio
nships': [{'Type': 'CHILD', 'Ids': ['b814c5cc-3c87-43e9-a04b-a2a3407e20f4', '9f1b08c3-df9f-479d-bec9-32ddcea0c095',
'406361d9-51b9-4989-867e-eadc5d7afe4a', 'a15cd582-cbe1-4b4d-9ec4-8683f850b425', '06d99f01-2187-46c9-90ec-e5aab35829
30', 'c5f1b5ac-b8a3-4aa3-90b7-4a530e3edec3', '265f332d-5fc9-44cd-8d41-274448fed14c', 'cab98dd5-41c9-4c33-8b87-850f3
efe6c50', 'c2798606-0a0d-4064-8b52-bc955fce8721', '92f7006f-f43e-4094-ac0a-80ef22d46b16', '685547aa-73ce-46c5-86e0-
e6537a20b09c', '5893365c-752f-4d43-ba11-5f29b28a4135', 'e66581ff-ab1b-4bd0-a4d5-5bc797971044', '7213668e-6817-4e7c-
a994-542599253bdd', '3107c5ef-a5c1-4d0e-a46a-04fb5901e582', 'c58a9b3e-be12-4b1e-ad65-b3084940d615', '32ed5907-f73d-
408f-ba16-a134c888cabf', '19b6475a-c42d-45a9-a37c-e9b2ae23ee7e', '030ae9fc-2969-48e0-8734-7abdba1c918c', '5f84c926-
a767-491c-bb5e-264444afa88a', '7f2e18e0-1c6e-4092-9bc7-43b3b4096c63', '1748fd60-f078-4685-89c6-1ee2d91df849', '8340
86a2-5b3f-4e1b-a705-65df49bab708', '749cbc2f-1516-44ec-9133-dac2c75fd3c4', 'ef3b33c8-35ae-4fc3-90e3-597640a3cb8e',
'462aa022-d1c8-4976-a45d-df688117e096', '468212c9-0a02-416a-9da3-645421de8641', '34e08f90-31a3-4e2b-b2e4-dd5870c1e4
```

Figure 10.3 – Amazon Textract JSON response

5.  We are writing a few lines of code to parse the Textract JSON response to extract all key-value pairs. This is defined as a `getformkeyvalue()` method for further reuse:

```
#Extract key values
# Iterate over elements in the document
from trp import Document
def getformkeyvalue(response):
    doc = Document(response)
    key_map = {}
    for page in doc.pages:
        # Print fields
        for field in page.form.fields:
            if field is None or field.key is None or
field.value is None:
                continue
```
```

```
                    key_map[field.key.text] = field.value.text
          return key_map
```

6. In the following code, we are calling the `getformkeyvalue()` method and printing the response:

```
get_form_keys = getformkeyvalue(response)
print(get_form_keys)
```

7. You can see next that Amazon Textract has extracted all key-value pairs and checkbox items from our URLA-1003 form:

```
{'Refinance': 'NOT_SELECTED', 'Purchase': 'SELECTED', 'Construction-Permanent': 'NOT_SELECTED', 'Secondary Residenc
e': 'NOT_SELECTED', 'FHA': 'NOT_SELECTED', 'Unmarried (includes single, divorced, widowed)': 'NOT_SELECTED', 'Const
ruction': 'NOT_SELECTED', 'Fee Simple': 'SELECTED', 'USDA/Rural Housing Service': 'NOT_SELECTED', 'Conventional': '
SELECTED', 'made': 'NOT_SELECTED', 'Fixed Rate': 'SELECTED', 'to be made': 'NOT_SELECTED', "Borrower's Name (includ
e Jr. or Sr. if applicable)": 'Alejandro Rosalez', 'Agency Case Number': '8562057552458', 'VA': 'NOT_SELECTED', 'To
tal (a+b)': '$', 'Rent': 'NOT_SELECTED', 'Amount Existing Liens': '$', '(a) Present Value of Lot': '$', 'Home Phone
(incl. area code)': '(888)555-0101', 'Lender Case Number': '58432890-8', 'Original Cost': '$', 'Subject Property Ad
dress (street, city, state, & ZIP)': '28777 Amos Lock, Markfurt, HI 71418', '(b) Cost of Improvements': '$', 'Socia
l Security Number': '518-85-7136', 'DOB (mm/dd/yyyy)': '05/31/1982', 'Yrs. School': '16', 'Own': 'NOT_SELECTED', 'P
rimary Residence': 'SELECTED', 'Interest Rate': '2.89 %', 'Amount': '$ 450,000.00', 'Title will be held in what Nam
```

Figure 10.4 – Amazon Textract FORM JSON response – URLA-1003

8. Moreover, we are running Textract queries on the URLA-1003 document to get answers to specific questions. As given in the following code block, we are passing in the query as `What is the preferred language?`, and Amazon Textract returns `English` as the response from our URLA-1003 document:

```
response = None
with open(documentName, 'rb') as document:
    imageBytes = bytearray(document.read())
    # Call Textract
    response = client.analyze_document(
        Document={'Bytes': imageBytes},
        # new QUERIES Feature Type for Textract Queries
        # We could add additional Feature Types like
FORMS and/or TABLES
        # FeatureTypes=["QUERIES", "FORMS", "TABLS"],
        FeatureTypes=["QUERIES"],
        QueriesConfig={
            "Queries": [{
                "Text": "What is the preferre language?",
                "Alias": "LANGUAGE"
            }
```

```
        ]
    })
print(response)
```

In addition to the loan application form, most often, additional supporting documents are processed. Let's check one such document, `Bank-Statement`, for accurate extraction with Amazon Textract:

1. First, let's check our `Bank-Statement` form. You can run the following command to check the image:

    ```
    # Document
    documentName = "Bank-Statement.jpg"
    display(Image(filename=documentName))
    ```

2. A sample bank statement is shown here:

Page 1 of 5   04/04/2022
DC   1090001004290

Example Inc. Credit Union

999-99-99-99 16769 3 C 001 11 S 66 002
ALEJANDRO ROSALEZ
400 KOZEY LIGHT, WEBERBURGH, HI 29922

Your consolidated statement

For 04/04/2022

Contact us

 example.com    (858) LLL-0101 or
(858) 555-0101

Do more with digital banking

Bank without having to leave home. Check your account balances, make transfers, pay bills and deposit checks with your mobile device. If you are not enrolled in digital banking, it only takes a minute Gel started today at example.com/U.

Example Bank, Member FDIC. To learn more, visit **example.com/ABCXYZ** ©2020 Example Inc. Credit Union.

If you are traveling outside of the USA and have concerns about accessing your account while you are traveling, please contact your Branch Banker or call us at 858-LLL-0101.

Summary of your accounts

ACCOUNT NAME	ACCOUNT NUMBER	BALANCE ($)	DETAILS ON
CHECKING	003525801543	5,657.47	page 1
Total checking and money market savings accounts		$5,657.47	
SAVINGS	352580154336	53,578.24	page 3
Total savings accounts		$53,578.24	

Figure 10.5 – Bank statement sample

3. Now, we are sending the Bank-Statement document to the calltextract() method defined previously. Then, we are passing in the JSON response of the calltextract() method to the getformkeyvalue() method, and extracting all form elements from the sample paystub document:

```
response= calltextract(documentName)
get_form_keys = getformkeyvalue(response)
print(get_form_keys)
```

4. You can check the sample output from Bank-Statement here:

```
Image loaded Bank-Statement.jpg
{'Your new balance as of 06/17/2020': '= $5,657.47', 'Checks': '– 1,177.33', 'Deposits, credits and interest': '+
3,124.75', 'Your previous balance as of 04/04/2022': '$41,982.42', 'Average Posted Balance in Statement Cycle': '$6
5,360.07', 'Other withdrawals, debits and service charges': '– 567.18', 'DATE': '06/09', 'CHECK #': '985026', 'AMOU
NT ($)': '150.00', 'Total checks': '= $701.39', 'Example Inc.': 'Credit Union', 'DC': '1090001004290', 'CHECKING':
'003525801543'}
```

Figure 10.6 – Bank statement JSON FORM output

During mortgage application processing, we often receive additional supporting documents such as pay stubs. We will use the following code for accurate extraction of pay stubs:

5. First, let's check our Bank-Statement Paystub form. You can run the following command to check the image:

```
# Document
documentName = "Paystub.jpg"
display(Image(filename=documentName))
```

A sample pay stub is given as follows:

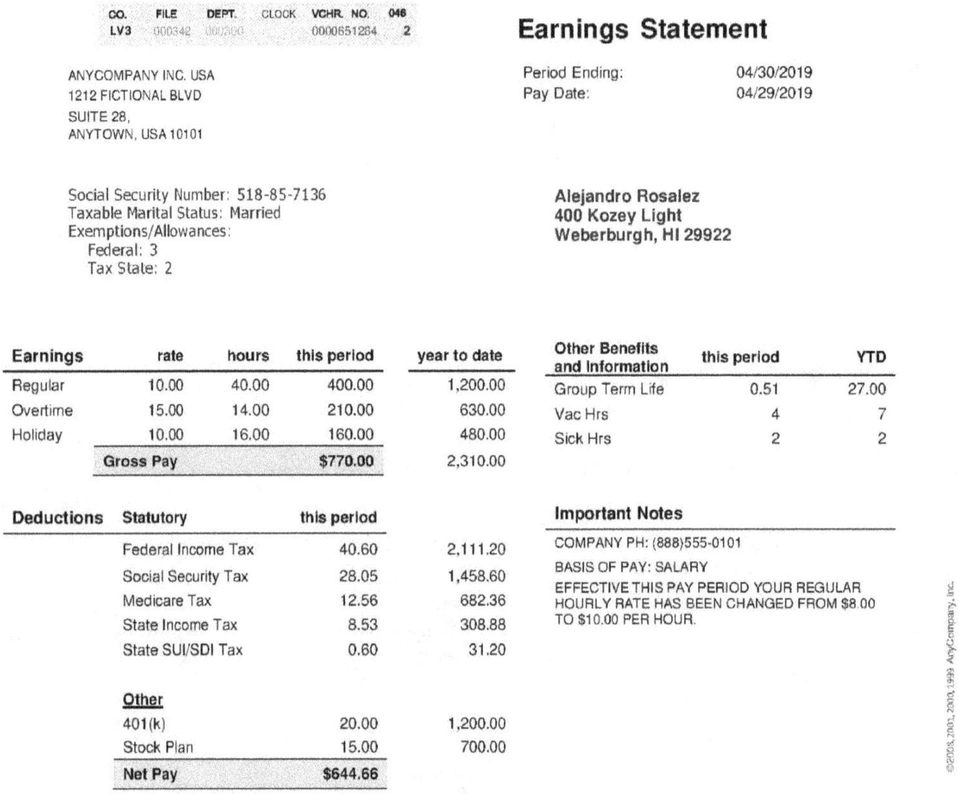

CO.	FILE	DEPT.	CLOCK	VCHR. NO.	046
LV3	000342	000300		0000651284	2

Earnings Statement

ANYCOMPANY INC. USA
1212 FICTIONAL BLVD
SUITE 28,
ANYTOWN, USA 10101

Period Ending:	04/30/2019
Pay Date:	04/29/2019

Social Security Number: 518-85-7136
Taxable Marital Status: Married
Exemptions/Allowances:
   Federal: 3
   Tax State: 2

Alejandro Rosalez
400 Kozey Light
Weberburgh, HI 29922

Earnings	rate	hours	this period	year to date
Regular	10.00	40.00	400.00	1,200.00
Overtime	15.00	14.00	210.00	630.00
Holiday	10.00	16.00	160.00	480.00
Gross Pay			**$770.00**	2,310.00

Other Benefits and Information	this period	YTD
Group Term Life	0.51	27.00
Vac Hrs	4	7
Sick Hrs	2	2

Deductions	Statutory	this period	
	Federal Income Tax	40.60	2,111.20
	Social Security Tax	28.05	1,458.60
	Medicare Tax	12.56	682.36
	State Income Tax	8.53	308.88
	State SUI/SDI Tax	0.60	31.20
	Other		
	401(k)	20.00	1,200.00
	Stock Plan	15.00	700.00
	Net Pay	**$644.66**	

Important Notes

COMPANY PH: (888)555-0101

BASIS OF PAY: SALARY

EFFECTIVE THIS PAY PERIOD YOUR REGULAR
HOURLY RATE HAS BEEN CHANGED FROM $8.00
TO $10.00 PER HOUR.

©2006, 2000, 2000, 1999 AnyCompany, Inc.

Figure 10.7 – Pay stub sample document

6. Now, we are sending the `Paystub` document to the `calltextract()` method defined previously. Then, we are passing in the JSON response of the `calltextract()` method to the `getformkeyvalue()` method, and extracting all `form` elements from the sample paystub document:

```
response= calltextract(documentName)
get_form_keys = getformkeyvalue(response)
print(get_form_keys)
```

7. You can check the sample output from `Paystub` as follows:

```
Image loaded Paystub.jpg
{'Advice number:': '650101270', 'Pay date:': '11/29/2019', 'Pay Date:': '04/29/2019', 'account number': 'xxxxx5967
', 'transit ABA': 'XXXX XXXX', 'Federal:': '3', 'Period Ending:': '04/30/2019', 'Social Security Number:': '518-85-
7136', 'Tax State:': '2', 'amount': '$644.66', 'Taxable Marital Status:': 'Married', 'VCHR NO.': '0000651284', 'FIL
E': '000342', 'Deposited to the account of': 'Alejandro Rosalez', 'CO.': 'LV3', 'COMPANY PH:': '(888)555-0101', 'DE
PT.': '000300', 'Net Pay': '$644.66', 'BASIS OF PAY:': 'SALARY', 'Gross Pay': '$770.00'}
```

Figure 10.8 – Paystub JSON FORM output

During mortgage application processing, we often receive additional supporting documents such as Form 1099-DIV. We will use the following code for the accurate extraction of the Form 1099-DIV type of document:

1. First, let's check our `1099-DIV` form. You can run the following command to check the image:

    ```
    # Document
    documentName = "1099-DIV.jpg"
    display(Image(filename=documentName))
    ```

    A sample of `1099-DIV` is shown here:

JANUS
PO Box 55932
Boston, MA 02205-5932

1-800-525-3713

RECIPIENT'S Name, Street Address (including apt. no.), City, State, and Zip Code

J   VAND

LITTLE ROCK AR 72227-2122

Tax Year 2015

Copy B For Recipient | Keep For Your Records

This is important tax information and is being furnished to the Internal Revenue Service. If you are required to file a return, a negligence penalty or other sanction may be imposed on you if this income is taxable and the IRS determines that it has not been reported.

Department of the Treasury - Internal Revenue Service

RECIPIENT'S Identification Number	8478

Form 1099-DIV • Dividends and Distributions (OMB No. 1545-0110)

1a. Total Ordinary Dividends	1b. Qualified Dividends	2a. Total Capital Gain Distributions	3. Nondividend Distributions	4. Federal Income Tax Withheld	6. Foreign Tax Paid
Fund Name: **JANUS GROWTH AND INCOME FUND D SHARES**			Account No.: **40-2003**	Payer's Fed. ID No.: **84-116**	
52.17	52.17	326	0.00	0.00	0.00
Fund Name: **JANUS FUND D SHARES**			Account No.: **42-2003**	Payer's Fed. ID No.: **84-059**	
32.84	87.06	796	0.00	0.00	0.00
Fund Name: **JANUS TWENTY FUND D SHARES**			Account No.: **43-2003**	Payer's Fed. ID No.: **84-097**	
90.13	90.13	503	0.00	0.00	0.00
Fund Name: **JANUS ENTERPRISE FUND D SHARES**			Account No.: **50-2003**	Payer's Fed. ID No.: **84-120**	
26.04	26.04	38	0.00	0.00	0.00

Figure 10.9 – Sample Form 1099

2. Now, we are sending the `1099-DIV` document to the `calltextract()` method defined previously. Then, we are passing in the JSON response of the `calltextract()` method to the `getformkeyvalue()` method, and extracting all `form` elements from the sample form 1099 document:

    ```
    response= calltextract(documentName)
    get_form_keys = getformkeyvalue(response)
    print(get_form_keys)
    ```

3. You can check the sample output from 1099-DIV as follows:

```
Image loaded 1099-DIV.jpg
{'Account No.:': '40-2003', "Payer's Fed. ID No.:": '84-120', "RECIPIENT'S Identification Number": '8478', 'RECIPIE
NTS Name, Street Address (including apt. no.), City, State, and Zip Code': 'J VAND LITTLE ROCK AR 72227-2122', 'Fun
d Name:': 'JANUS ENTERPRISE FUND D SHARES'}
```

Figure 10.10 – Textract JSON response – Form 1099

Now, let's see how we can extract additional ID documents such as a US passport:

4. We are using the Passport.pdf document. We pass in raw bytes from the document to Textract's analyze_id method. The AnalyzeID API can extract and normalize the fields from the passport type of document:

```
# Document
documentName = "Passport.pdf"
with open(documentName, 'rb') as document:
    imageBytes = bytearray(document.read())
response = client.analyze_id(
    DocumentPages=[{"Bytes":imageBytes}]
)
print(json.dumps(response, indent=2))
```

5. Next, you can check the JSON response from the Passport document:

```
{
  "IdentityDocuments": [
    {
      "DocumentIndex": 1,
      "IdentityDocumentFields": [
        {
          "Type": {
            "Text": "FIRST_NAME"
          },
          "ValueDetection": {
            "Text": "ROSALEZ",
            "Confidence": 98.47076416015625
          }
        },
        {
          "Type": {
            "Text": "LAST_NAME"
          },
          "ValueDetection": {
```

Figure 10.11 – Textract JSON response – Passport document

Now we will show you another way to extract values from the Paystub document using Amazon Textract Queries. In the preceding code, we showed you how you can call Amazon Textract's FORMS API to get a JSON response and can use your parsing logic to get answers to exact areas such as *gross pay* and *net pay*. But can we extract the values of *gross pay* and *net pay* directly from Paystub without doing any post-processing? By using the following code with Textract Queries, we can extract the values.

In the following code, we are getting raw bytes from `Paystub` and passing them as a blob to the `analyze_document()` method with FeatureType as `"QUERIES"`. We pass in the questions such as `What is the gross pay?` and `What is the Net pay?` as `QueriesConfig` to the `analyze_document()` method:

```
# Document
documentName = "Paystub.jpg"
response = None
with open(documentName, 'rb') as document:
    imageBytes = bytearray(document.read())
    # Call Textract
    response = client.analyze_document(
        Document={'Bytes': imageBytes},
        # new QUERIES Feature Type for Textract Queries
        # We could add additional Feature Types like FORMS and/
or TABLES
        # FeatureTypes=["QUERIES", "FORMS", "TABLES"],
        FeatureTypes=["QUERIES"],
        QueriesConfig={
            "Queries": [{
                "Text": "What is the gross pay?",
                "Alias": "GROSS_PAY"
            },
            {
                "Text": "What is the net pay?",
                "Alias": "NET_PAY"
            }
        ]
        })
    print(response)
```

You can see the Textract Queries feature accurately extracted Gross and Net pay from our sample document. There are additional ways to extract entities from documents with AWS AI services, such as Amazon Comprehend. Let's now check how we can leverage Amazon Comprehend to extract fields from documents for mortgage application processing.

Extraction with Comprehend

We can use Amazon Comprehend for further extraction of elements for mortgage loan application processing. We can leverage Comprehend predefined entities to detect entities such as `"Person"`, `"Date"`, `"Quantity"`, and more from any documents. For example, we can leverage `"Mortgage Note"` types of unstructured documents and can extract predefined entities from this document. At times, we have a need to extract entities, those that are custom entities or key business terms specific to the use case and business requirements. To extract those entities, we can leverage Amazon Comprehend's **custom entity recognition** feature. For example, in the same `"Mortgage Note"` document type, we have a requirement to extract `"Payment"` information. We can leverage Amazon Comprehend to extract these elements. In the following figure, you can see a high-level flow to extract entities from supporting documents for mortgage loan application processing with Amazon Comprehend:

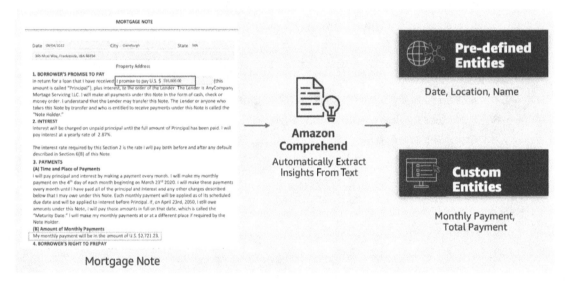

Figure 10.12 – Comprehend entities and custom entities extraction

Now, let's see how we can implement the same for further extraction with Amazon Comprehend:

1. We are using `Paystub` as a sample document. We want to extract all predefined entities from this `Paystub` sample. First, we need to extract all text from this pay stub for further processing. To extract text from this document, we are calling the `calltextract()` method defined previously:

    ```
    documentName = "Paystub.jpg"
    response= calltextract(documentName)
    print(response)
    ```

2. Now, let's use the following code to print all the lines extracted by Amazon Textract:

```
# Print detected text
text = ""
for item in response["Blocks"]:
    if item["BlockType"] == "LINE":
        print ('\033[94m' + item["Text"] + '\033[0m')
        text = text + " " + item["Text"]
```

You can check the output lines as printed here:

```
CO.
FILE
DEPT.
CLOCK VCHR NO.
046
LV3
000342
000300
0000651284
2
Earnings Statement
ANYCOMPANY INC. USA
Period Ending:
04/30/2019
1212 FICTIONAL BLVD
Pay Date:
04/29/2019
SUITE 28,
ANYTOWN, USA 10101
```

Figure 10.13 – Textract lines – Paystub document

3. After all the lines are extracted, we are calling the Amazon Comprehend **detect entities** feature to extract all predefined entities from Paystub. We are using Amazon Comprehend's **Real-time analysis** on AWS Console:

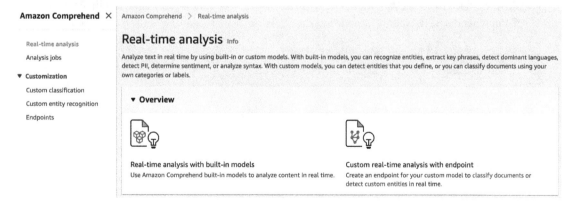

Figure 10.14 – Amazon Comprehend on AWS Console

4. We are passing all text in *Step 3* to Amazon Comprehend for further extraction. You can see next that our entities of interest (such as **Person**, **Organization**, **Location**, and **Date**) are extracted with a confidence score from the `Paystub` document type:

04/29/2019	Date	0.99+
ANYCOMPANY INC.	Organization	0.99+
USA	Location	0.96
Alejandro Rosalez	Person	0.99+
CO. FILE DEPT. CLOCK	Organization	0.96
ANYCOMPANY INC.	Organization	0.99+
ANYTOWN, USA 10101	Location	0.86

Figure 10.15 – Comprehend predefined entities – Paystub

In the preceding code examples, we learned how to leverage AWS AI services for the extraction of elements from documents for mortgage application processing.

Document enrichment for mortgage application processing

In this section, we will show you an example of document enrichment for mortgage application processing. We will use the `Passport` type of document. We will use Amazon Comprehend to detect sensitive information from the document. Let's dive into the code:

1. First, we need to extract all lines from the `Passport` document to be analyzed by Amazon Comprehend. Use the following code to extract all lines from the sample `Passport` document:

```
documentName = "Passport.pdf"
response= calltextract(documentName)
# Print detected text
text = ""
for item in response["Blocks"]:
    if item["BlockType"] == "LINE":
        print ('\033[94m' +  item["Text"] + '\033[0m')
        text = text + " " + item["Text"]
```

You can check all the lines extracted from the document here:

```
Image loaded Passport.pdf
PASSPORT
UNITED STATES OF AMERICA
PASSEPORT
Type/Type/Tipo Code/Code/Codigo
Passport No./ No. du Passeport / No. de Passporte
PASSPORTE
P USA
918268822
Surname / Nom / Appelidos
USA
Alejandro
Given Names / Prénoms / Nombres
Rosalez
Nationality / /Nationalité/Nacionalidad
UNITED STATES OF AMERICA
Date of birth / Date de naissance / Fecha de nacimiento
15 Apr 1990
Place of birth / Lieu de naissance / Lugar de nacimiento
Sex / Sexe / Sexo
Texas, U.S.A.
M
Date of issue / Date de délivrance / Fecha de expedición
Authority / Autorité / Autoridad
29 Jan 2009
United States
Date of expiration / Date d'expiration / Fecha de caducidad
31 Jan 2029
Department of State
Endorsements / Mentions Speciales / Anotaciones
SEE PAGE 27
P<USAROSALEZ<<ALEJANDRO<<<<<<<<<<<<<<<<<<<<<<<
349587345029USA1209381M2394820392344059< < 834983457
```

Figure 10.16 – Textract lines – Passport document

2. Then, we are passing all the lines extracted by Amazon Textract in *Step 2* to Amazon Comprehend. We are using Amazon Comprehend's **Real-time analysis** on AWS Console. We are also passing in the text and then clicking on **Analyze**:

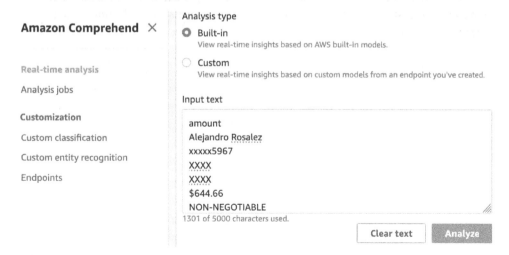

Figure 10.17 – Comprehend on AWS Console

3. Click on the **PII** tab to check the detected PII entities from the sample `Passport` document with confidence scores:

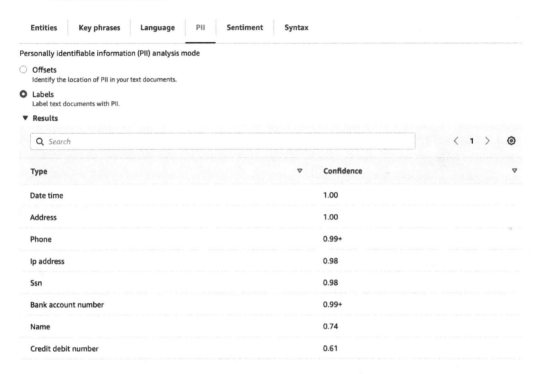

Type	Confidence
Date time	1.00
Address	1.00
Phone	0.99+
Ip address	0.98
Ssn	0.98
Bank account number	0.99+
Name	0.74
Credit debit number	0.61

Figure 10.18 – Comprehend PII on AWS Console

4. After the PII elements are extracted by Amazon Comprehend, you can leverage the **bounding box** feature of Amazon Comprehend and Textract to map the elements on the document. You can have your custom code for the redaction of these elements on the document, and you can see the sample high-level flow here:

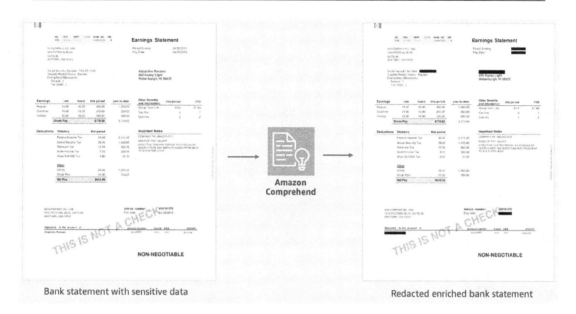

Figure 10.19 – Comprehend PII redaction

Now we have learned about how we can leverage various stages of the IDP pipeline for mortgage application processing use cases. IDP not only applies to mortgage application processing but can apply to all financial industry use cases where we have a need to process documents. Now let's look at some of those use cases.

Understanding the mortgage processing review and verification stage of the IDP pipeline

Further, you can use the review and verification stage of the IDP pipeline for mortgage application processing use cases, as mentioned in *Chapter 6, Review and Verification of Intelligent Document Processing*. At a high level, we have a completeness check and accuracy check to process a document in the review and validation stage of the IDP pipeline. The user defines their business rules to be checked against the extracted elements from a document. For example, for mortgage processing, we want to check whether all required documents are submitted or not. Moreover, we want to make sure that all fields are accurately filled in on the URLA-1003 form. We can also do cross-document verification. For example, we can check **Name** and **DOB** across identity documents and loan application documents. At times, we have a requirement to include human beings to review and correct the elements during the claims processing use case. I recommend you check the *Learning document review process with human loop* section of *Chapter 6, Review and Verification of Intelligent Document Processing*.

Understanding financial services use cases for document processing

Most banks or financial institutions still continue to process account opening forms, income verification documents, credit card applications, and so on, through manual paper-based forms. The extent of digital transactions for these activities has significantly increased over the years but paper-based forms (or digital images of paper forms) and manual processing of those forms continue to be significantly prevalent, especially where the banking ecosystem is getting connected to other parts of society and the business ecosystem. The ability to automatically and accurately read, store, and manage data from unstructured text will allow for complete digitization and automation of those processes, which will improve accuracy, reduce processing costs and risks, improve associate engagement, and enable digitization of broader financial transactions.

KYC standards are used within the financial services industry to verify customers and their financial and risk profiles. The requirements and/or standards are broadly applicable in the financial services industry, including banking, brokerage, and investments. Various regulatory bodies (for example, the **Securities and Exchange Commission (SEC)**, US **Financial Crimes Enforcement Network (FinCEN)**, and **Financial Industry Regulatory Authority (FINRA)**) have defined specific compliance requirements that organizations need to follow. The most important part of the KYC requirements is to establish proof of the customer's identity to ensure financial transactions are not performed by bad actors and that financial transactions can be tracked to the appropriate customers if needed in case of illegal activities such as money laundering. To establish proof of a customer's identity, customers are usually required to submit copies of unstructured text-based documents such as their passport, driver's license, bank or credit card statements, utility bills, address proof, and so on. Financial institutions typically use a significant amount of manual and human-based processing to perform the verification and maintenance activities, as most data is captured through paper, emails, and images, and they are not easily integrated across various systems, both internally and externally. In the last 10 to 20 years, significant progress has been made in digitizing the activities but the ability to accurately read, parse, and verify information in an automated manner will allow for a significant impact on costs, accuracy, and associate engagement, and the reduction of financial loss for the financial services industry. Major financial institutions have reported spending up to $500 million annually on KYC verification and compliance.

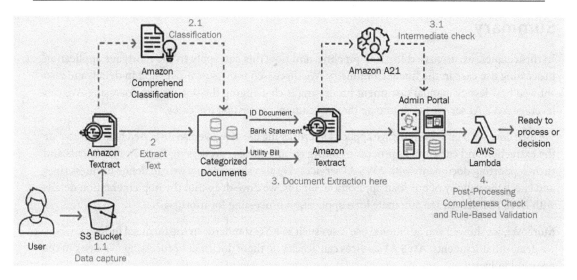

Figure 10.20 – Reference architecture for a KYC use case

To automate KYC document processing, we can leverage IDP with AWS AI services, as shown in the preceding reference architecture.

1. We store documents such as ID documents, bank statements, utility bills, and more in Amazon S3.

2. Then, we use Amazon Comprehend classification to classify documents for further processing.

3. After documents are categorized, we process ID documents such as passports, and driver's licenses through the Amazon Textract Analyze ID API, bank statements with the Textract FORM API, and bills with the Textract Invoice API.

4. We recommend having an option for human review with Amazon A2I.

Also note that we are dealing with PII information for document processing, and the main task of KYC is to check a person's identity. So, we cannot redact the required PII information for verification purposes. I recommend checking the following references to design a secure architecture. Once we have all the extracted information, we can use serverless compute for automated review and verification of information. For example, you want to verify that the ID document and utility bill belong to the same person. This is a reference architecture that we used to automate the KYC process.

These are some of the examples of document processing in the financial industry, which can be accelerated and automated with the help of AWS AI services such as Amazon Textract, Amazon Comprehend, and the AWS scalable, secure infrastructure.

Summary

In this chapter, we discussed the IDP pipeline, and how this can apply to the mortgage application processing use case in the financial industry. We discussed mortgage processing in detail, and also showed how legacy manual document processing is challenging the document processing. We can leverage AWS AI services to automate these document processing use cases.

We then dove deep into the mortgage application processing use case. We also discussed how to automate the extraction and enrichment process of mortgage application processing for loan documents and their supporting documents with AWS AI services. We also showed how we can define business rules and check the validity of our loan application. Finally, we dove deep into the implementation details with AWS IDP services to automate loan application processing for mortgages.

Moreover, we showed you additional use cases such as KYC standards in the financial industry, where we deal with documents. AWS AI services can accelerate these document processing use cases in the financial industry.

This brings us to the end of this book, *Intelligent Document Processing with AWS AI Services*. I hope that you understood the challenges in document processing, and how AWS AI services can help automate the document workflow. Now you can take the knowledge of the document processing pipeline to various use cases across industries. It is with a heavy heart that we bid "adieu" to you. We hope that you had as much fun reading this book as we had writing it.

References:

- *Document Splitter*: `https://aws.amazon.com/blogs/machine-learning/intelligentlysplit-multi-form-document-packages-with-amazon-textract-and-amazoncomprehend/`.

- *Security in AWS*: `https://docs.aws.amazon.com/whitepapers/latest/introductionaws-security/introduction-aws-security.pdf`

- *Security in AWS*: `https://docs.aws.amazon.com/textract/latest/dg/security.html`

Index

`Packt.com`

Subscribe to our online digital library for full access to over 7,000 books and videos, as well as industry leading tools to help you plan your personal development and advance your career. For more information, please visit our website.

Why subscribe?

- Spend less time learning and more time coding with practical eBooks and Videos from over 4,000 industry professionals

- Improve your learning with Skill Plans built especially for you

- Get a free eBook or video every month

- Fully searchable for easy access to vital information

- Copy and paste, print, and bookmark content

Did you know that Packt offers eBook versions of every book published, with PDF and ePub files available? You can upgrade to the eBook version at `packt.com` and as a print book customer, you are entitled to a discount on the eBook copy. Get in touch with us at `customercare@packtpub.com` for more details.

At `www.packt.com`, you can also read a collection of free technical articles, sign up for a range of free newsletters, and receive exclusive discounts and offers on Packt books and eBooks.

Other Books You May Enjoy

If you enjoyed this book, you may be interested in these other books by Packt:

Practical Deep Learning at Scale with MLflow

Yong Liu

ISBN: 9781803241333

- Understand MLOps and deep learning life cycle development
- Track deep learning models, code, data, parameters, and metrics
- Build, deploy, and run deep learning model pipelines anywhere
- Run hyperparameter optimization at scale to tune deep learning models
- Build production-grade multi-step deep learning inference pipelines
- Implement scalable deep learning explainability as a service
- Deploy deep learning batch and streaming inference services
- Ship practical NLP solutions from experimentation to production

Machine Learning on Kubernetes

Faisal Masood, Ross Brigoli

ISBN: 9781803241807

- Understand the different stages of a machine learning project
- Use open source software to build a machine learning platform on Kubernetes
- Implement a complete ML project using the machine learning platform presented in this book
- Improve on your organization's collaborative journey toward machine learning
- Discover how to use the platform as a data engineer, ML engineer, or data scientist
- Find out how to apply machine learning to solve real business problems

Packt is searching for authors like you

If you're interested in becoming an author for Packt, please visit `authors.packtpub.com` and apply today. We have worked with thousands of developers and tech professionals, just like you, to help them share their insight with the global tech community. You can make a general application, apply for a specific hot topic that we are recruiting an author for, or submit your own idea.

Share your thoughts

Now you've finished *Intelligent Document Processing with AWS AI/ML*, we'd love to hear your thoughts! Scan the QR code below to go straight to the Amazon review page for this book and share your feedback or leave a review on the site that you purchased it from.

`https://packt.link/r/1-801-81056-7`

Your review is important to us and the tech community and will help us make sure we're delivering excellent quality content.